Advanced Stair Stringer Layout Methods

By Greg Vanden Berge

Published by Greg Vanden Berge
Copyright 2012 Greg Vanden Berge

ISBN-13: 978-1478383574
ISBN-10: 1478383577

Greg's Books

Home Buyers Checklist
How To Build Strait Stairs
501 Contractor Tips
Simplified Stair Building
Guide For Hiring Contractors
Simplified Bracket Stair Building
Simplified Tile Floor Installation
Simplified House Inspection Checklist
Simplified Home Inspections

http://gregvandenberge.com

http://gregvan.com/book_deals.htm

http://gregvandenberge.com/contact.htm

Disclaimer

Greg Vanden Berge, and its owners, agents and employees, make no warranty respecting the accuracy or currency of any information in the content or pages of this book or any source document referenced herein or linked to herein. Use of this book is conditioned on the user's understanding and agreement that we shall not be liable, on any theory whatsoever, including but not limited to negligence, for any damages attributable to that use.

In no event shall Greg Vanden Berge, its owners, agents or employees be liable to you or anyone else for any decision made or action taken by in reliance on any content created by Greg Vanden Berge or other individuals, companies, corporations or parties. Greg Vanden Berge and its affiliates, agents, owners and employees shall not be liable to you or anyone else for any damages, including without limitation, consequential, special, incidental, indirect, or similar damages, even if advised of the possibility of such damages.

Your use of this book and all related rights and obligations, shall be governed by the laws of the United States of America, as if your use was a contract wholly entered into and wholly performed within the United States of America. Any legal action or proceeding with respect to this book or any matter related thereto may be brought exclusively in the courts of the United States of America. By using this book, you agree generally and unconditionally to the jurisdiction of the aforesaid courts and irrevocably waive any objection to such jurisdiction and venue.

Do not copy or distribute this book.

This manual contains materials protected under International and Federal Copyright laws and Treaties. Any unauthorized reprint or use of this material is prohibited.

Table of Contents

Basic Stringer Layout	4
Bottom Stringer Layout Methods	19
Concrete Foundations	19
On Top Of Bottom Landing	22
In Bottom Landing	23
Simple Landing Method	25
Landing Wall Method	31
On Top Of Landing	40
Hanger Method	43
Top Stringer Layout Methods	49
Beam And Ledger	49
Hanging Plywood Ledger	53
Attaching Stringer To Joist	61
In Floor Or Landing	68
Layout Tips And Rules	78
90 Degree Angles	78
Parallel Lines	79
Bottom Layout Riser Variations	80
Top Layout Riser Variations	85
Riser Variations (Thickness)	88
Using Landing Hangers	95
Landing Joist Variations	98
Landing Length Issues	99
Stinger Layout Mistakes	102
Stringer Cracks And Hangers	102

Basic Stringer Layout

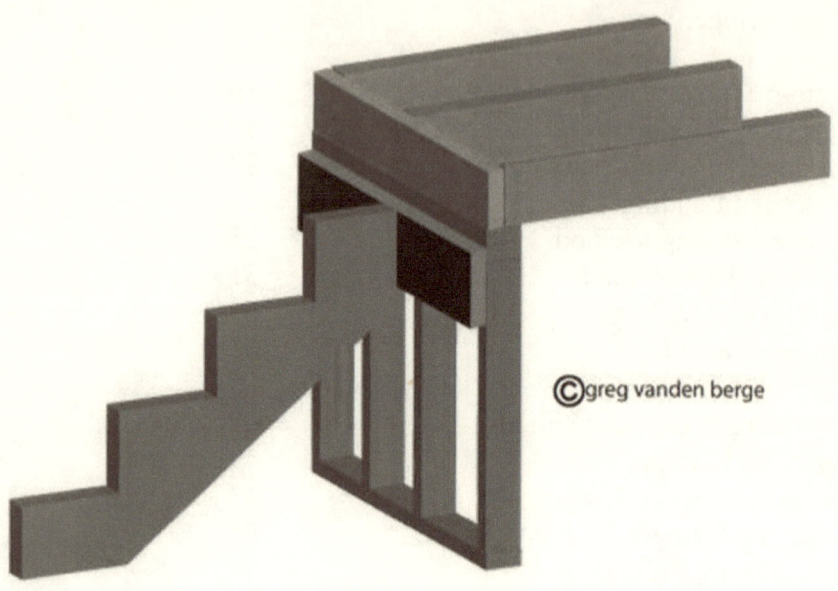

Before we get started, I'm going to provide you with a step-by-step set of instructions for laying out a basic stair stringer. Just in case you aren't familiar with laying out or building stairways.

Feel free to skip this section if you understand the basics.

This book won't provide you with step-by-step stair building instructions or how to figure out the rise and run. This book was written for construction workers, stair builders and everyone else interested in taking the next step to learn more about the fine art of stair building and stair stringer layout.

If you need stair building instructions use link below and check out the book," Simplified Stair Building."

http://stairs4u.com/stairbuildingbooks.htm

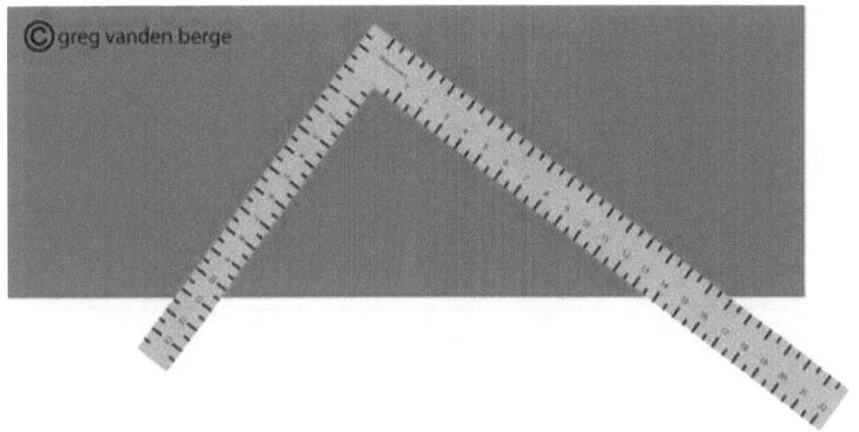

Step 1: Lay your framing square flat on top of the lumber you have chosen to use for your stringer as shown in picture above. You will be working from your right to your left. Position the framing square between 8 and 12 inches away from the right edge of the lumber.

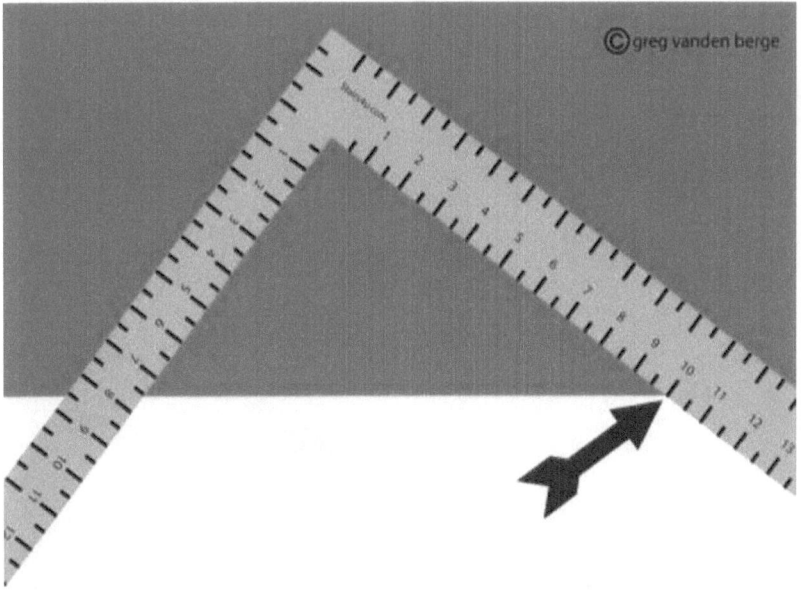

Step 2: Line up the individual step or tread measurement on the framing square with the edge of lumber.

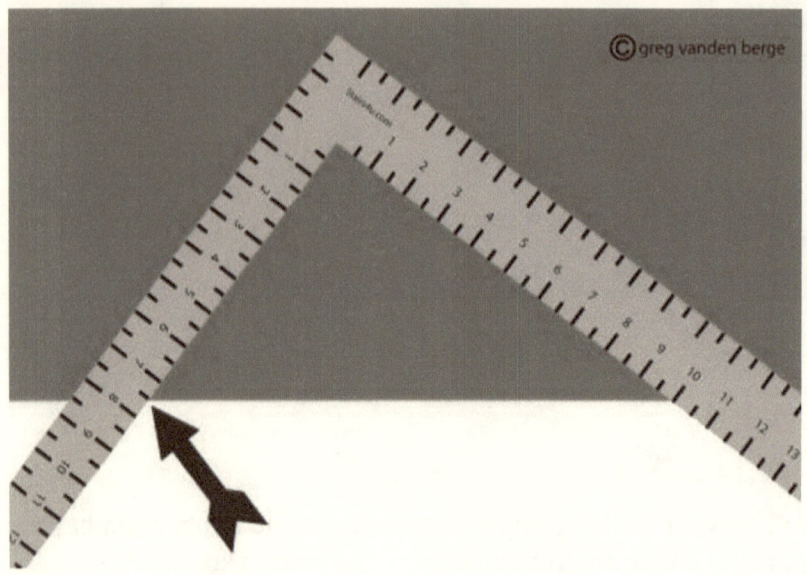

Step 3: Do the same on opposite side for the riser.

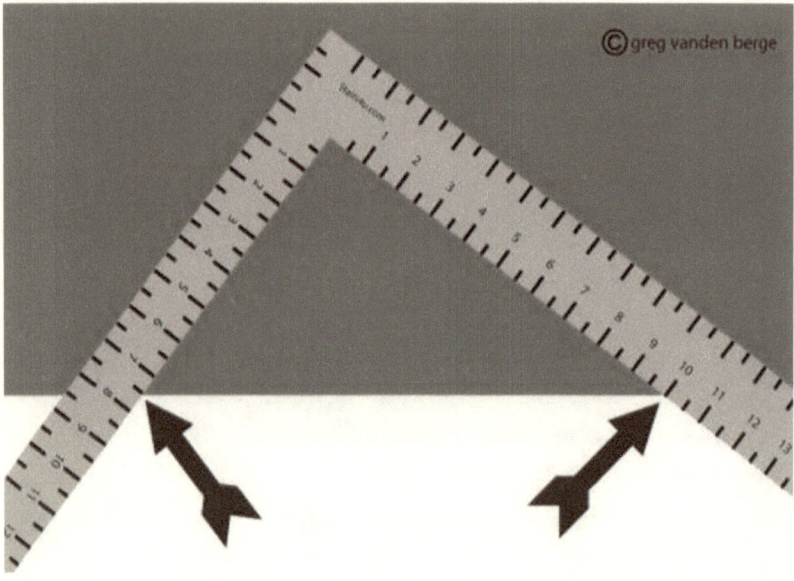

Step 4: In our example above and most of the examples throughout the book, we'll be using 7 1/2 inch risers and 10 inch treads. Don't make any marks on lumber until you have positioned your framing square correctly.

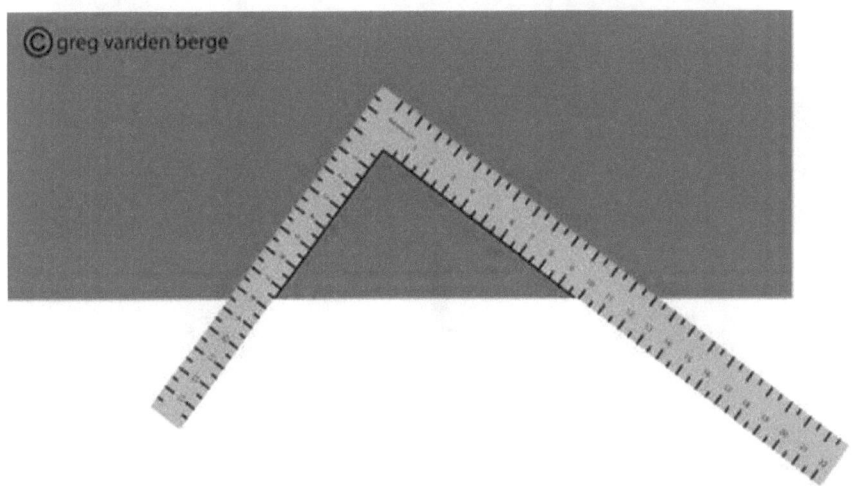

Step 5: Mark lumber as shown in picture above.

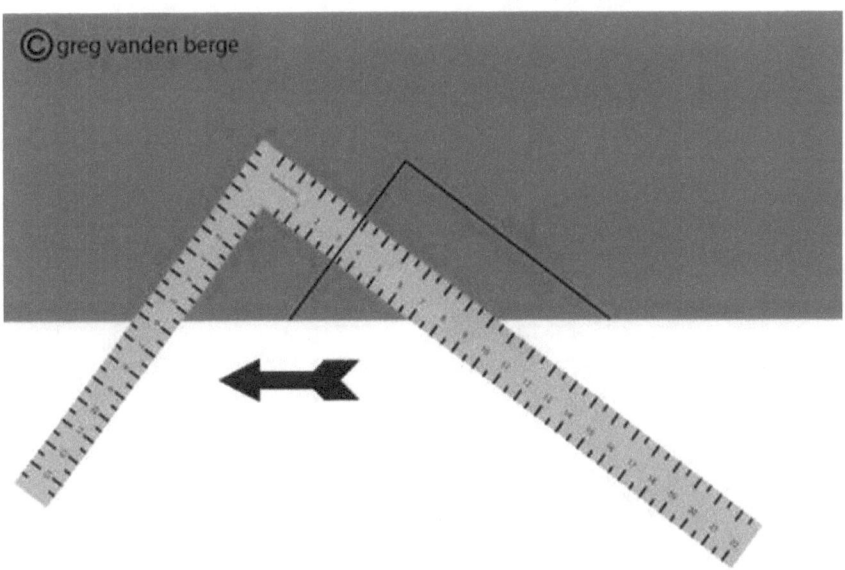

Step 6: After marking lumber slide framing square to the left.

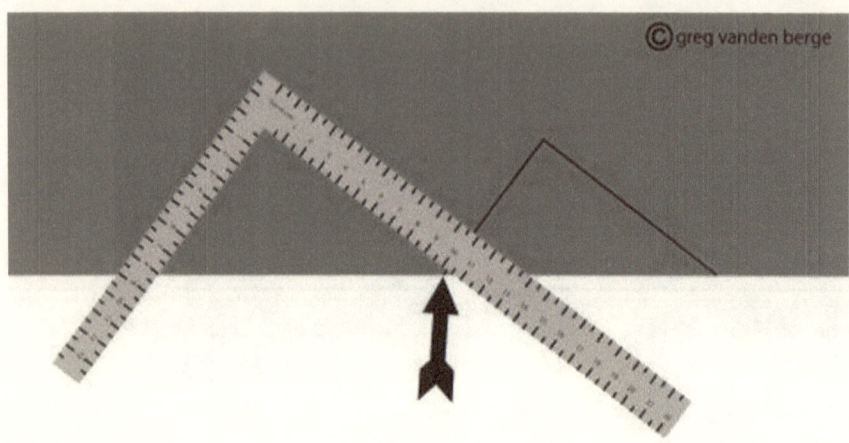

Step 7: The black arrow is pointing to the riser line you just marked, where it lines up with the edge of lumber.

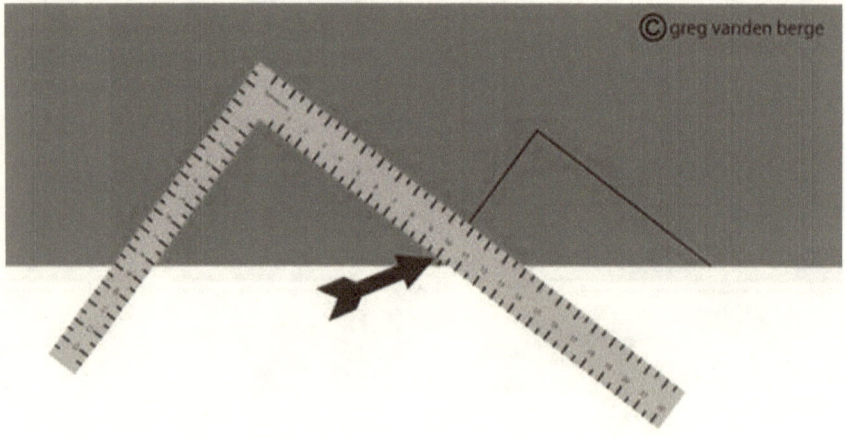

Step 8: The black arrow in the picture above is pointing to the 10 inch mark on the framing square that will represent each individual stair tread, throughout the rest of the book.

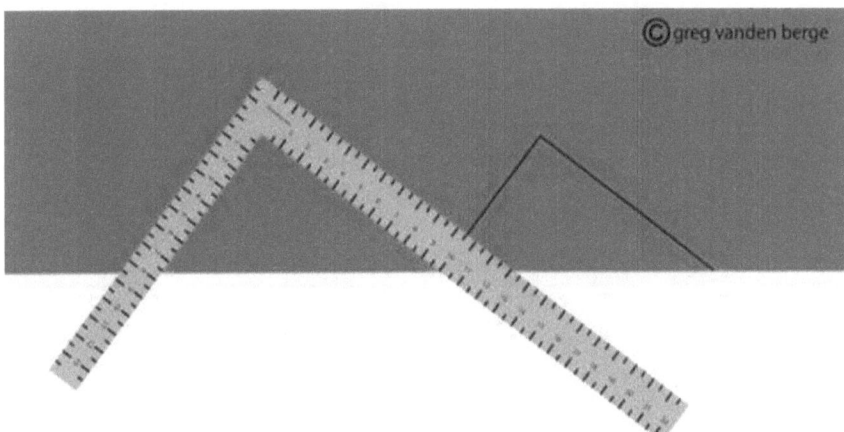

Step 9: Before you can mark the next tread and riser you will need to line the tread measurement mark on framing square up with the riser mark on lumber. Then you must re-position the framing square as shown in step four. Once this is done, mark lumber for your second tread and riser.

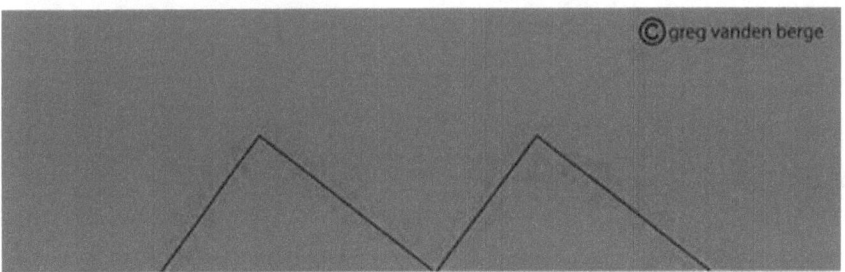

By now you should have something resembling the picture above.

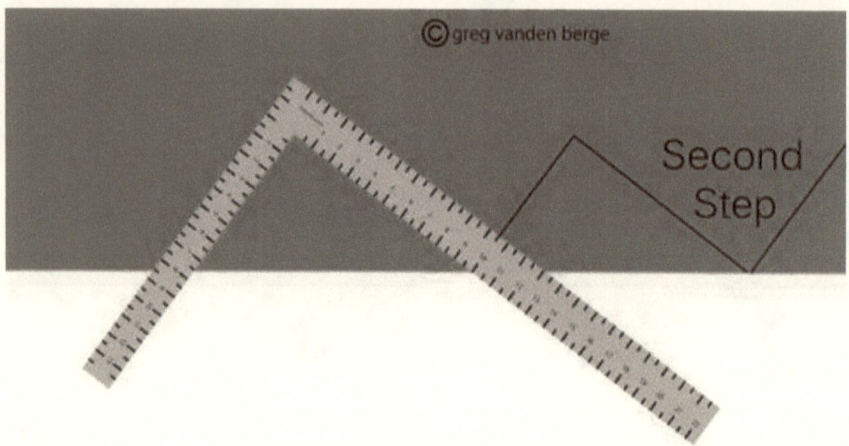

Step 10: In order to layout the next step simply repeat steps 6 through 9. You can layout as many treads and risers as you need to, depending upon the length of your stairway. In this example we will be laying out a three step stairway.

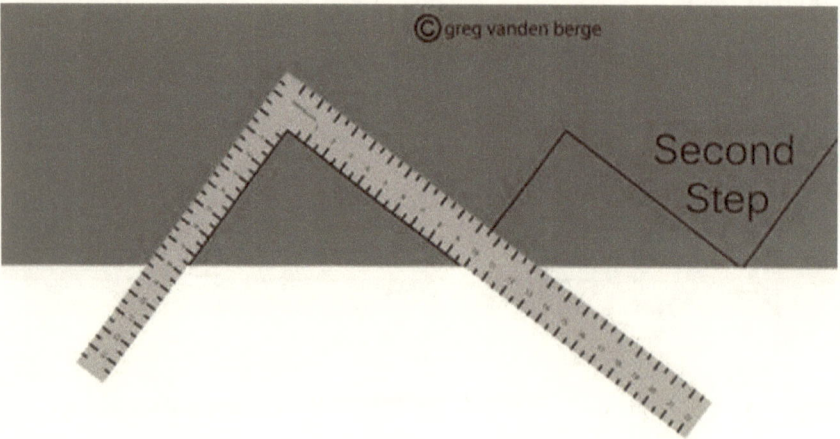

Step 11: After you have positioned the framing square correctly, mark the third step.

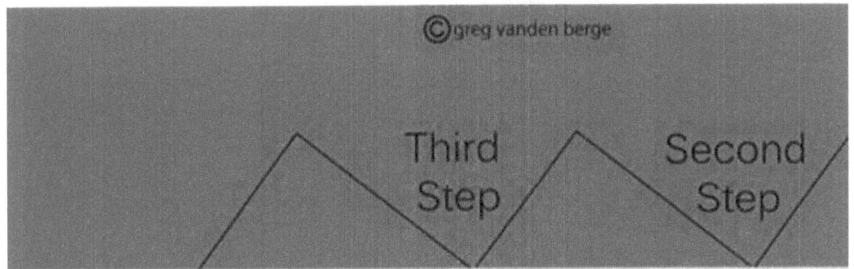

Step 12: By now the stair stringer should have three lines each representing your individual stair treads and risers. If you were laying out a 10 step stairway then you would need 10 lines representing each individual stair tread and riser.

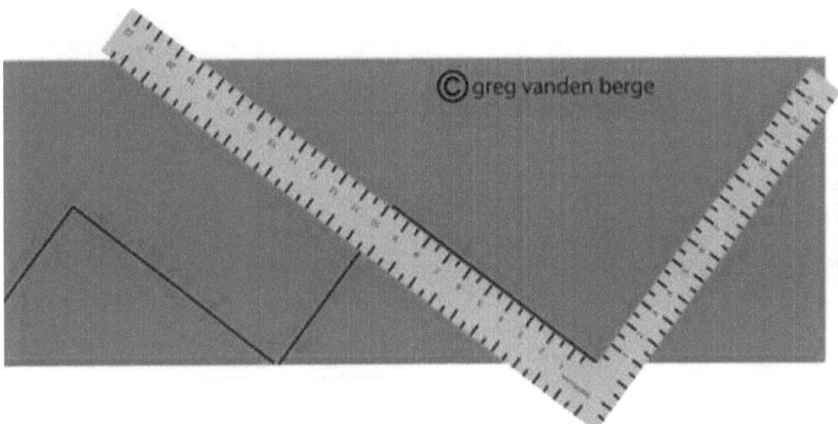

Step 13: Now it's time to layout the bottom of the stair stringer. Flip the framing square over and line the tread measurement on the framing square up with the point where the riser meets the stair tread as shown in picture above.

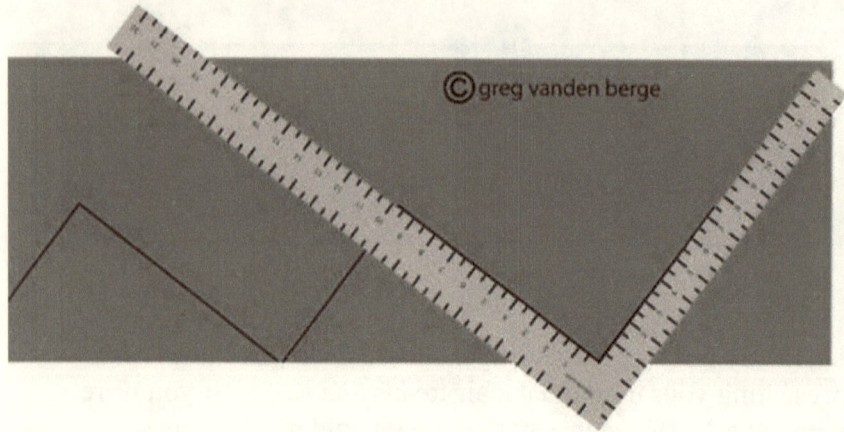

Step 14: After you've positioned the framing square correctly, feel free to mark the first or bottom riser.

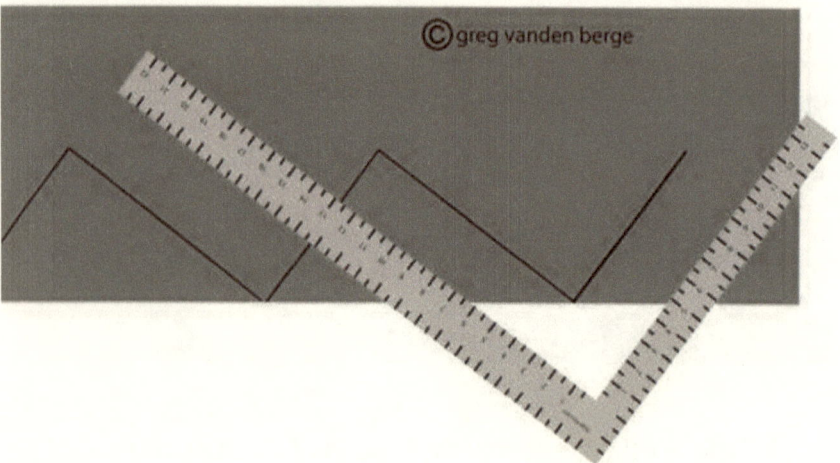

Step 15: You should end up with something looking like the picture above.

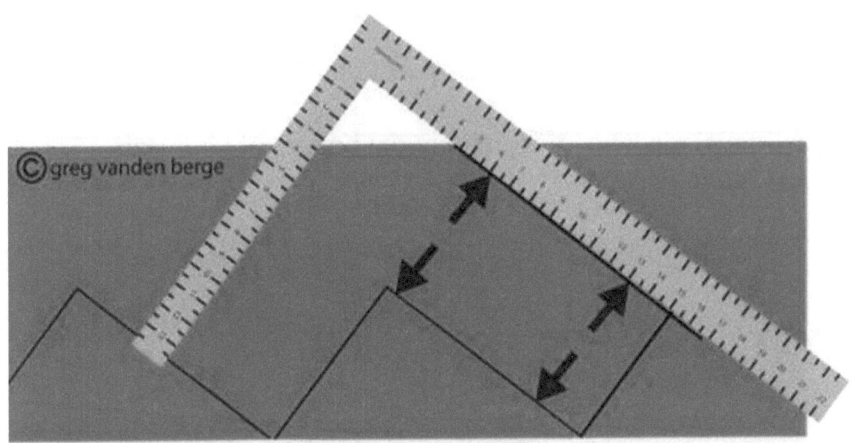

Step 16: Stringer bottom, overall riser measurements will vary from stairway to stairway. You will need to subtract the tread thickness and any other materials used to build the stairway, from the bottom of your stringer. For example, if I have a 7 1/2 inch over all riser and I'm going to use treads that are an inch and a half thick, then the distance between the arrows in the picture above would be 6 inches.

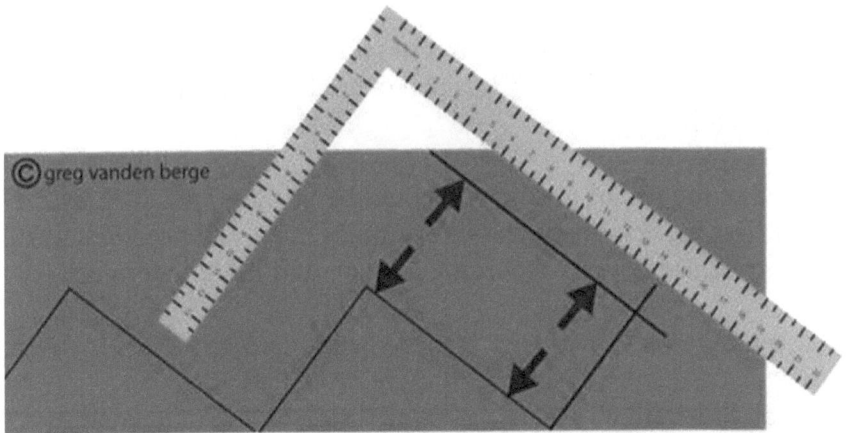

Step 17: If the stringer will be sitting on top of 2 x 4 treated lumber, separating it from a concrete building foundation then you will need to subtract an additional inch and a half. You would need to subtract 3 inches from your total bottom riser, giving you four and a half inches between the arrows shown in picture above.

See picture of stringer sitting on top of treated lumber on page 21.

For additional methods of laying out the bottom of your stair stringers, go to page 19.

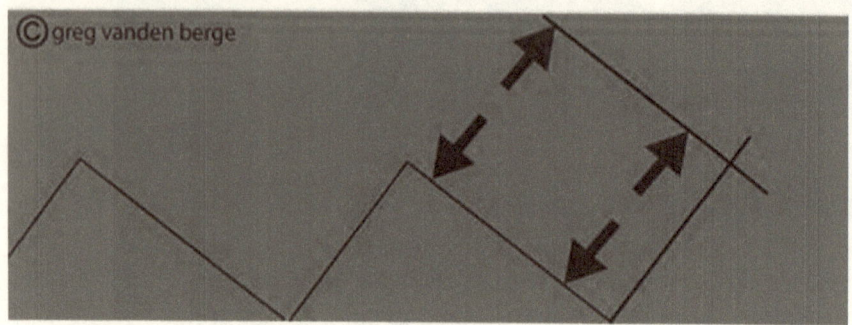

Step 18: If you're going to use 3/4" plywood treads and the stringer will be sitting on top of a wood framed landing or floor then you'll only need to subtract three quarters of an inch from the overall bottom riser measurement.

The distance in between the arrows in picture above would be 6 3/4" as long as we were using a 7 1/2 inch over all stair riser measurement.

Throughout the rest of this book we will be using a 7 1/2 inch over all stair riser measurement, in our examples. You will need to adjust your riser measurements according to your stair building project.

These measurements will vary from project to project.

To see an illustration of a stair stringer sitting on top of a wood framed landing on page 41.

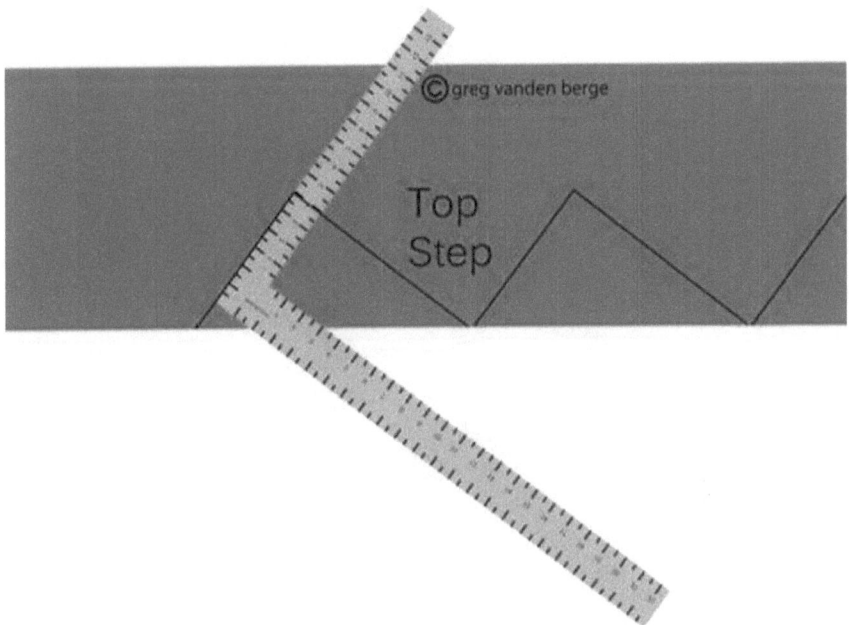

Step 19: Now it's time to layout the top of the stair stringer. In this example we will be laying out the top of this stringer, for a ledger. Line the framing square up with the last riser as shown in picture above.

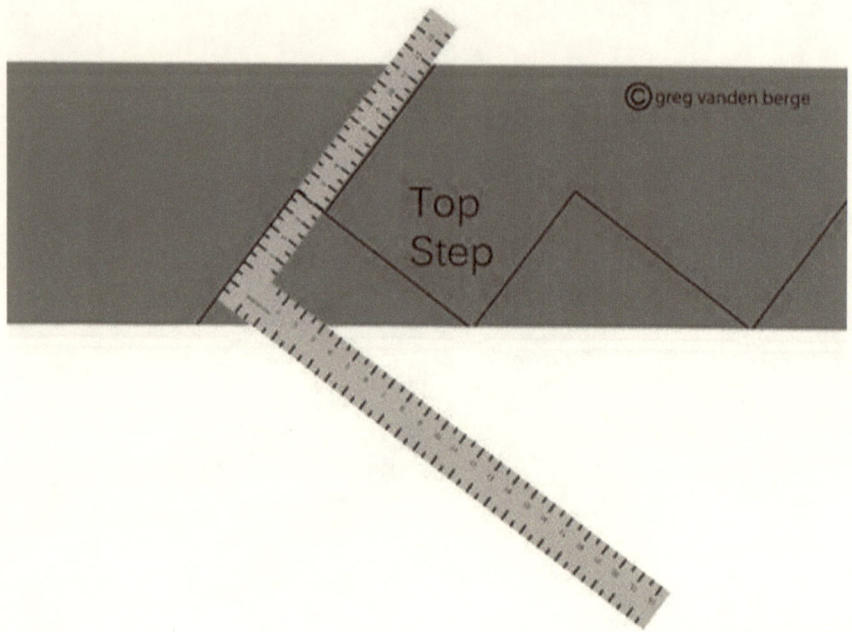

Step 20: Once you have the framing square properly positioned, you can mark the lumber. Make sure you use the inch and a half wide section of framing square and not the two inch side. The ledger in our example will be an inch and a half thick.

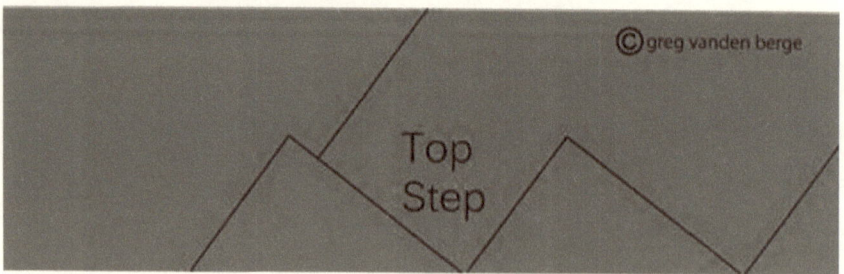

Step 21: You should end up with something like this.

Step 22: In this step I cut the top of the stair stringer and will be providing you with a few more illustrations to give you a better idea, why you're doing, what you need to do.

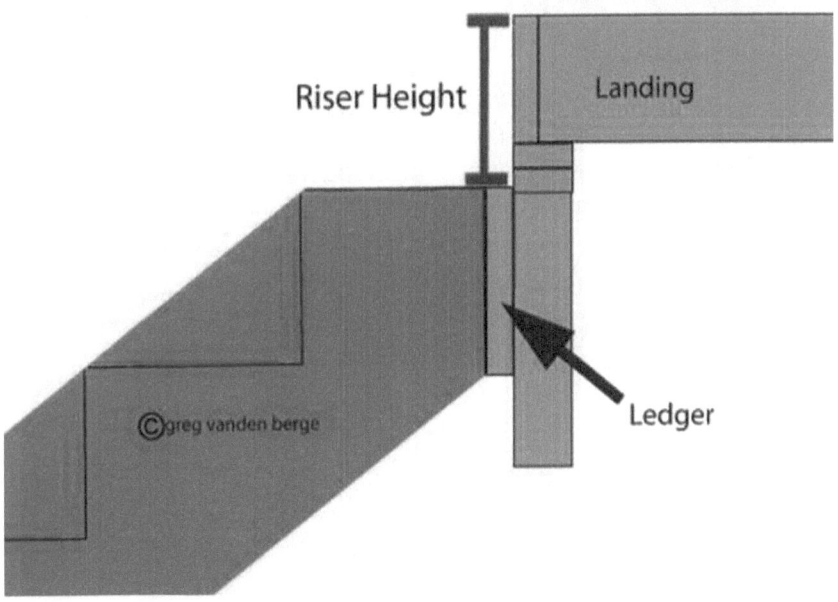

Step 23: In this illustration you can clearly see an inch and a half thick ledger and get a better idea why you subtracted the inch and a half from the last tread.

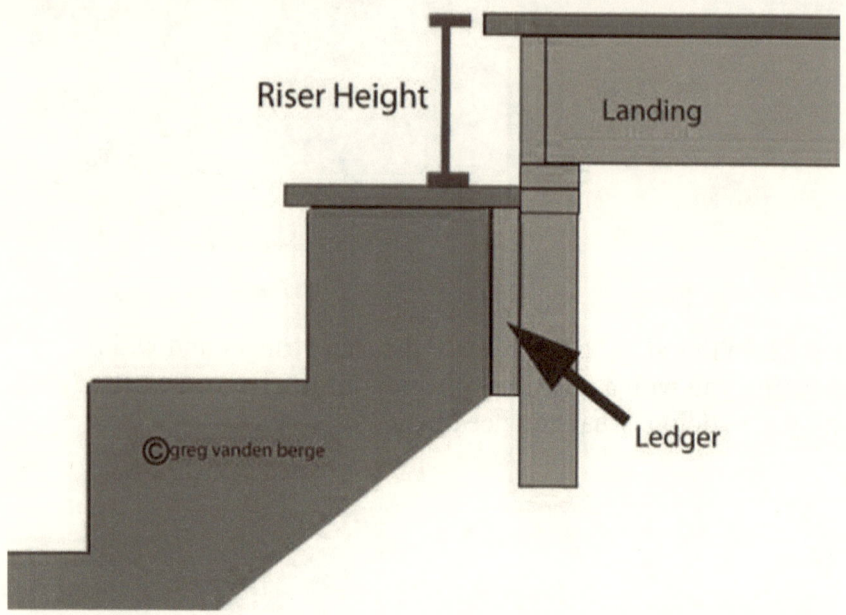

Step 24: After you install the top stair tread and landing sheathing, you will end up with something like this. As long as the materials you're going to use for your stair landing or floor are the same thickness as the stair treads.

If these materials are going to be different thicknesses then go to page 84 for further instructions.

Go to page 49 for additional ways to layout the top of your stair stringers.

Bottom Stringer Layout Methods

Concrete Foundations

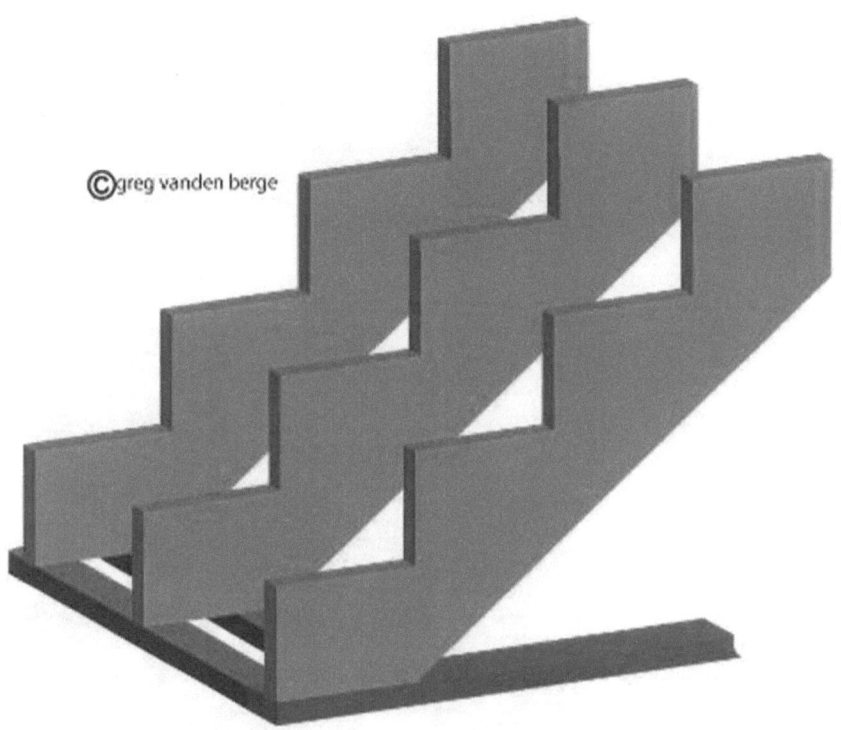

Most architects, engineers and building departments will not allow you to place construction standard lumber that isn't treated, directly on top of a concrete building foundation. However, there are exceptions to this rule. Feel free to contact your local building department for more information.

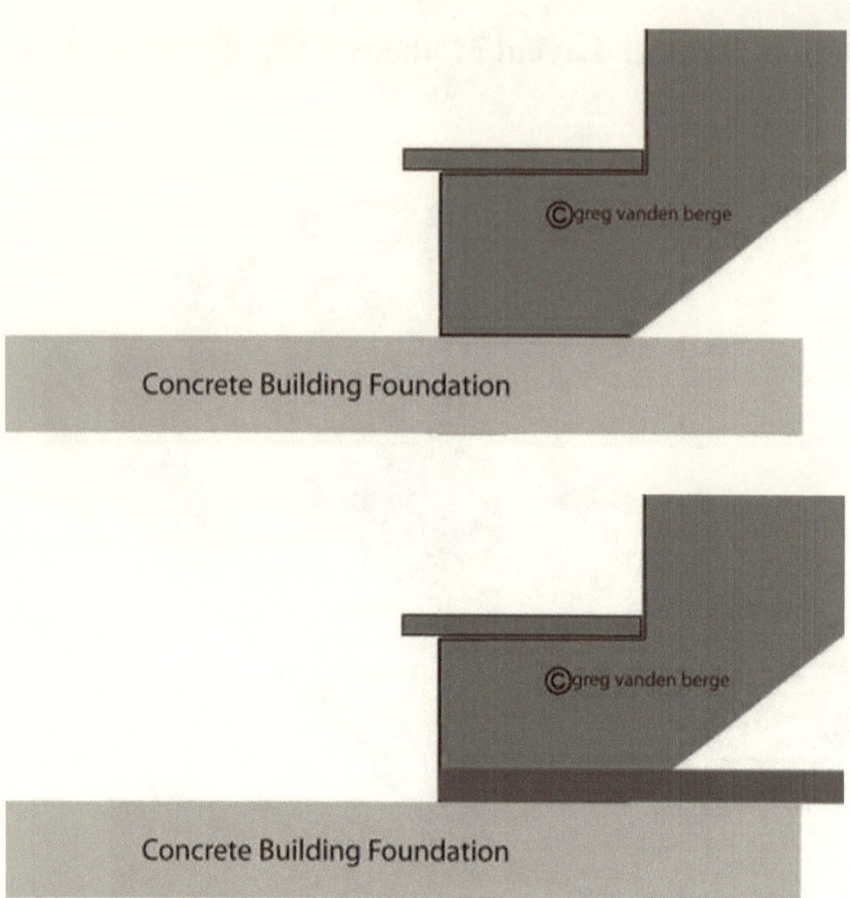

In between the concrete building foundation and stringer we have a green rectangle. This green rectangle represents a 2 x 4 piece of treated lumber that runs parallel or in the same direction as the stair stringers.

Concrete Building Foundation

In this illustration we're looking at the same side view of the stairway, except the treated 2 x 4s are running perpendicular to the stringers. Either one of these methods can be used under the right circumstances.

This is the most popular way to attach stair stringers to a concrete foundation.

For Layout Instructions Follow Steps 13 - 17

On Top Of Bottom Landing

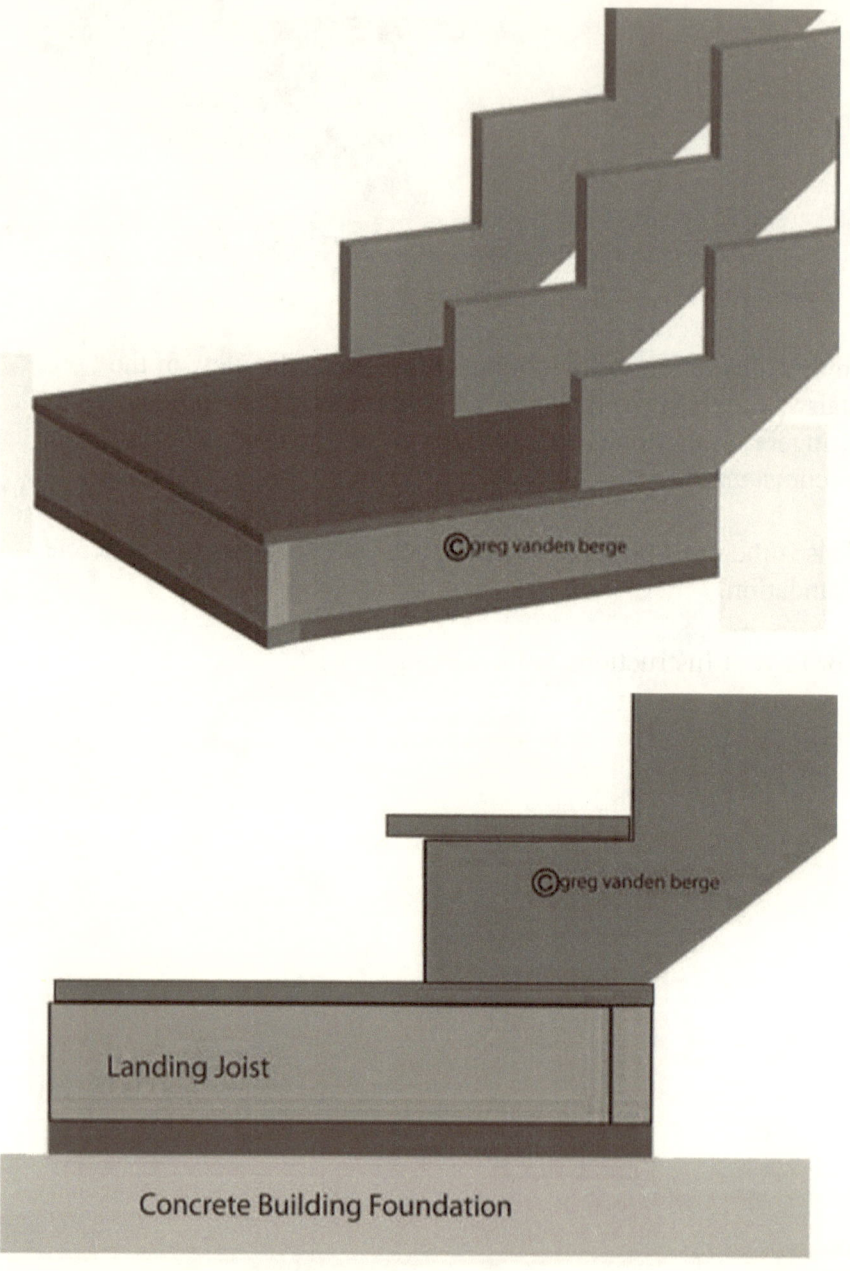

For Layout Instructions Follow <u>Steps 13 -16 and 18</u>

In Bottom Landing

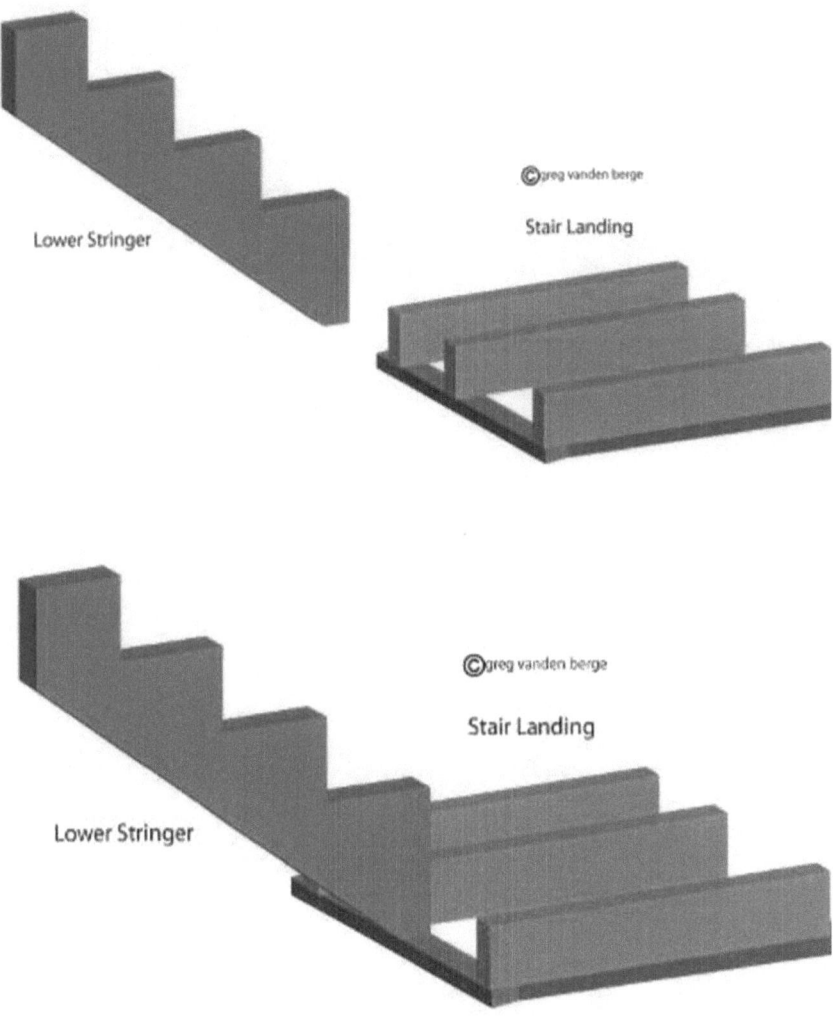

For Layout Instructions Follow Step 1-13

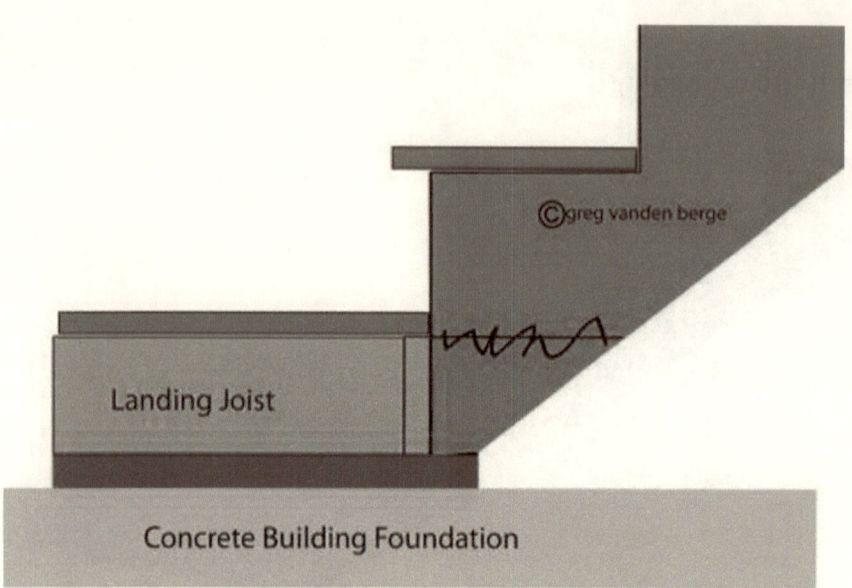

Simple Landing Method

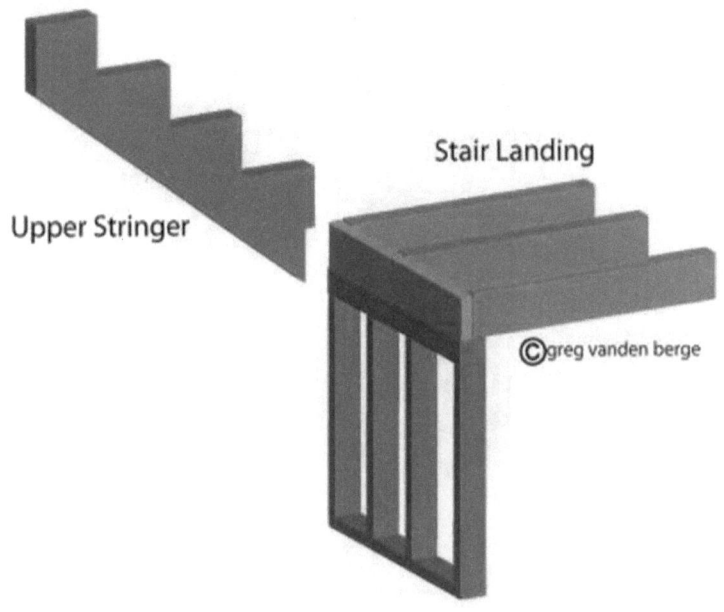

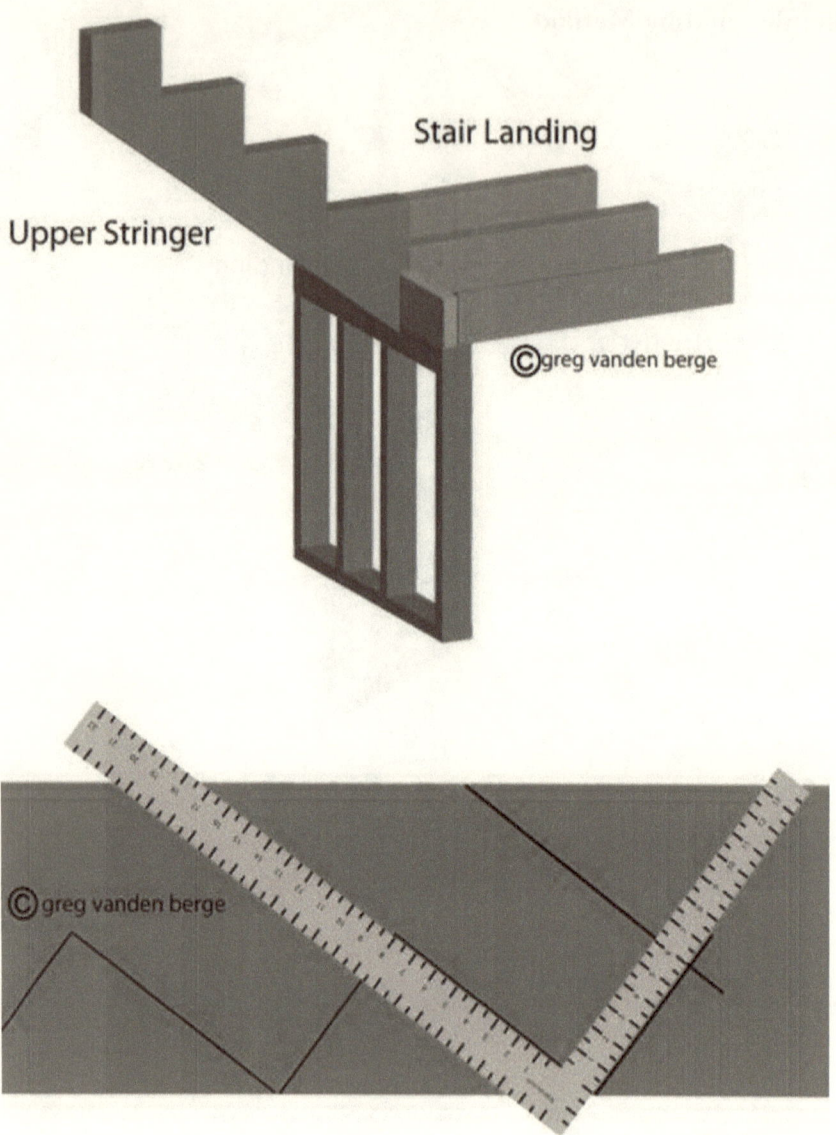

Step 1: Line the inch and a half side of your framing square up with bottom riser as shown in picture above.

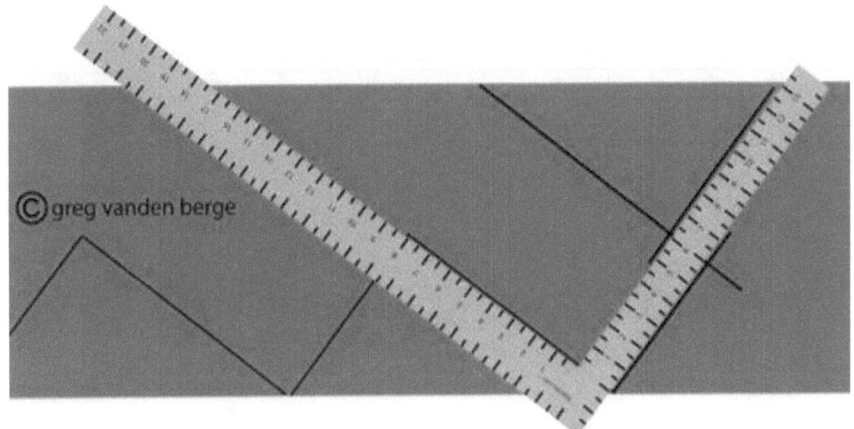

Step 2: After the framing square is properly positioned, mark lumber as shown in picture above.

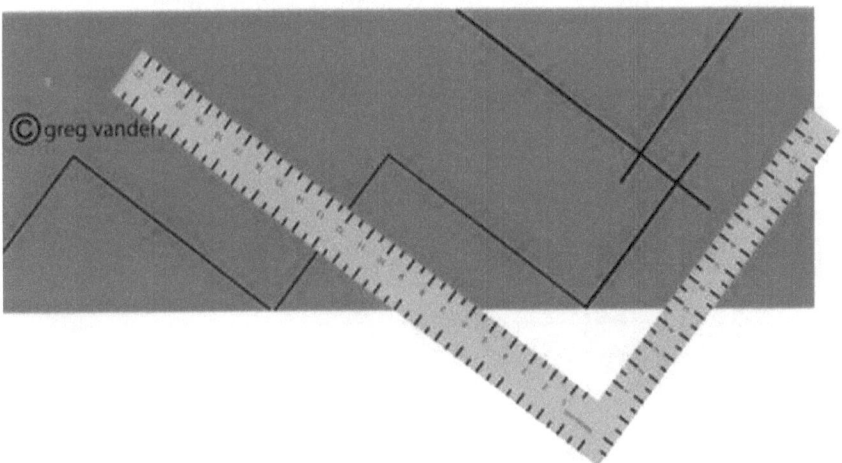

Step 3: You should end up with something like this.

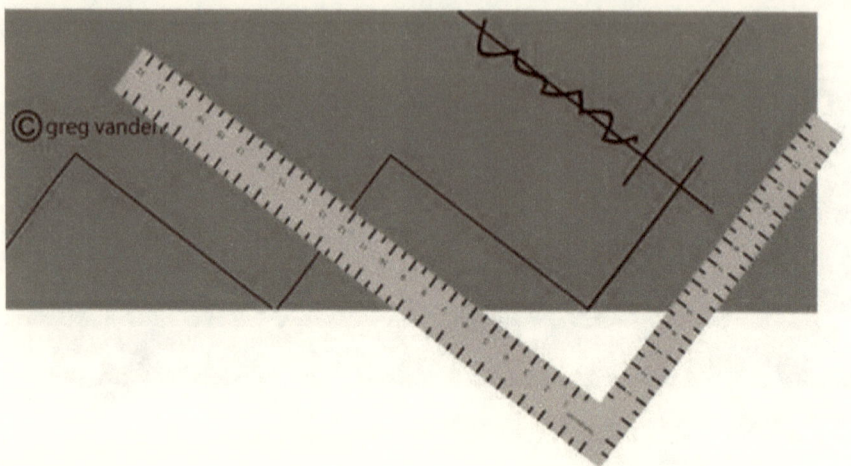

Step 4: Now this next part is extremely important. Make sure you cross out any lines you don't plan on cutting. I can't tell you how many times I cut one of these lines when I shouldn't have.

Step 5: The bottom of the stringer should look something like this after it's cut.

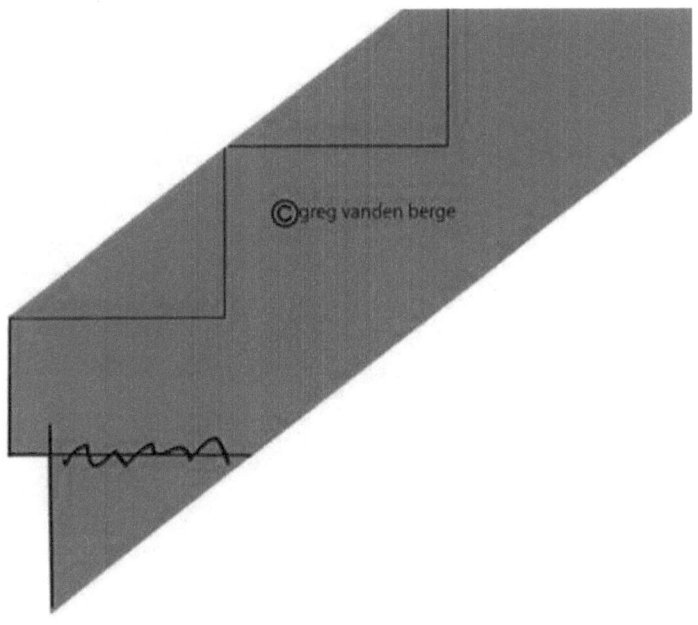

Step 6: I went ahead and flipped the stringer over, to give you a better idea what the stringer would look like, before attaching it to the landing.

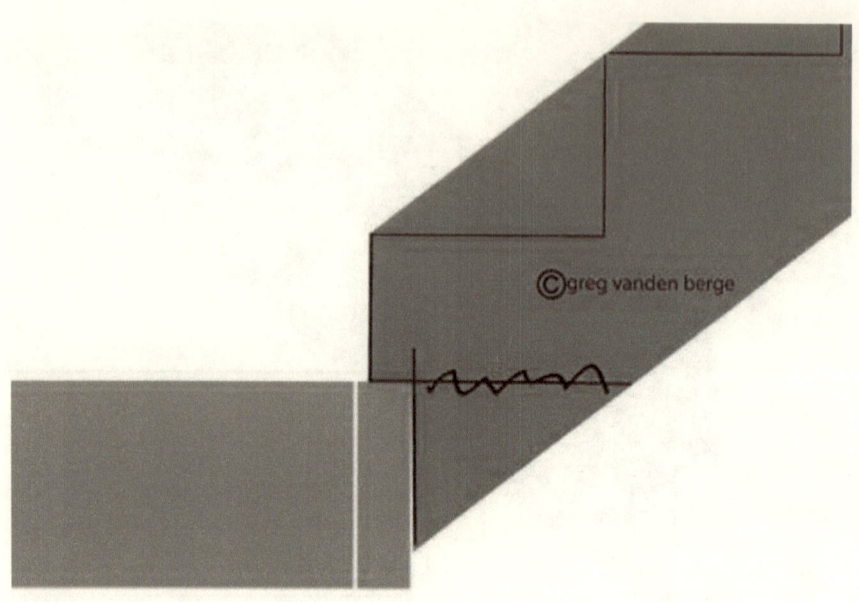

Step 7: This is what the stringer would look like after it was attached to the landing. I've probably installed more than a thousand stairs using this method.

Landing Wall Method

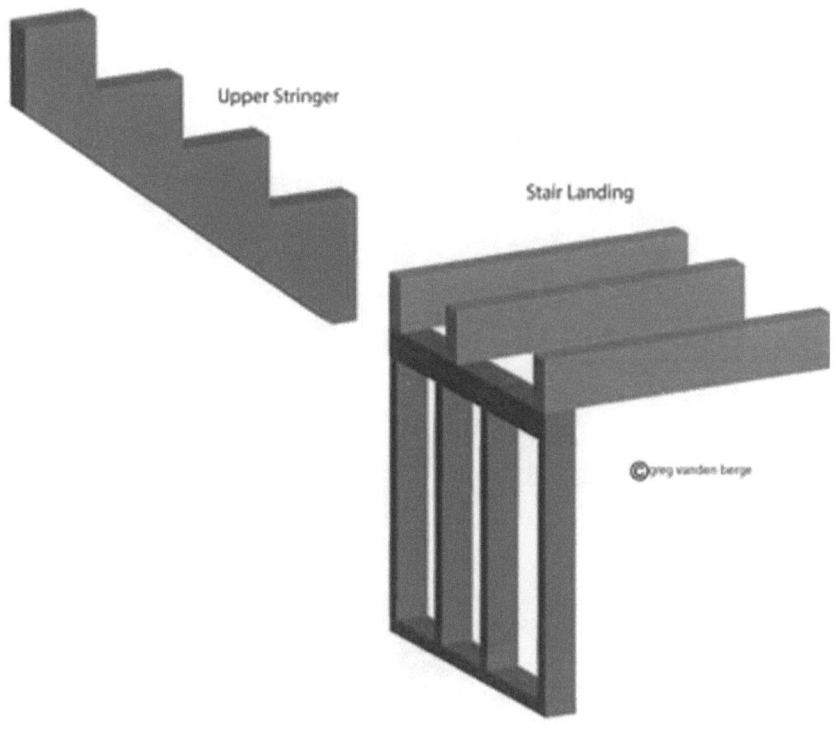

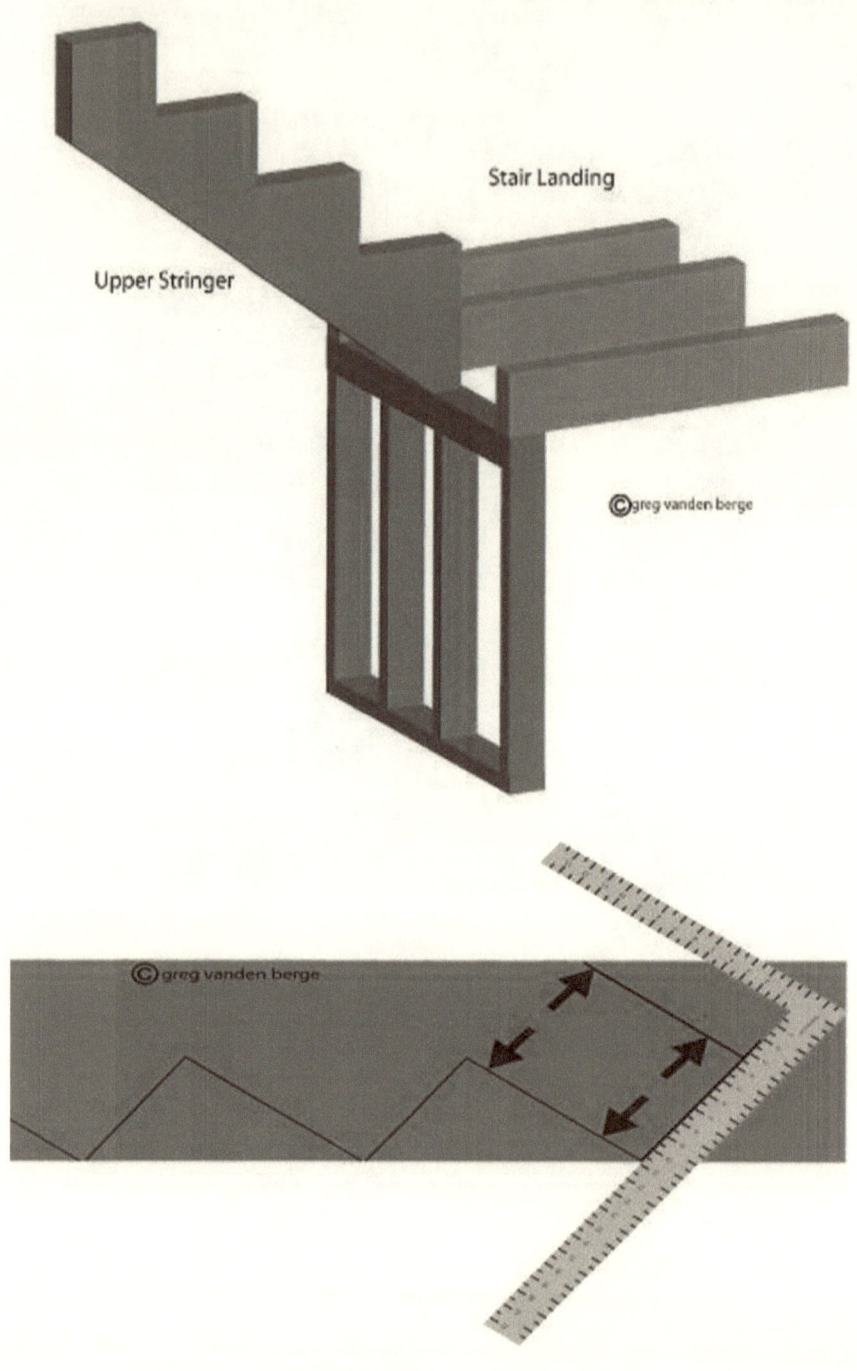

Step 1: This time we're going to use the 2 inch side of the framing square, while lining it up with our bottom riser as shown in picture above. Refer back to <u>Step 16</u>, if you need help figuring out the distance between black arrows.

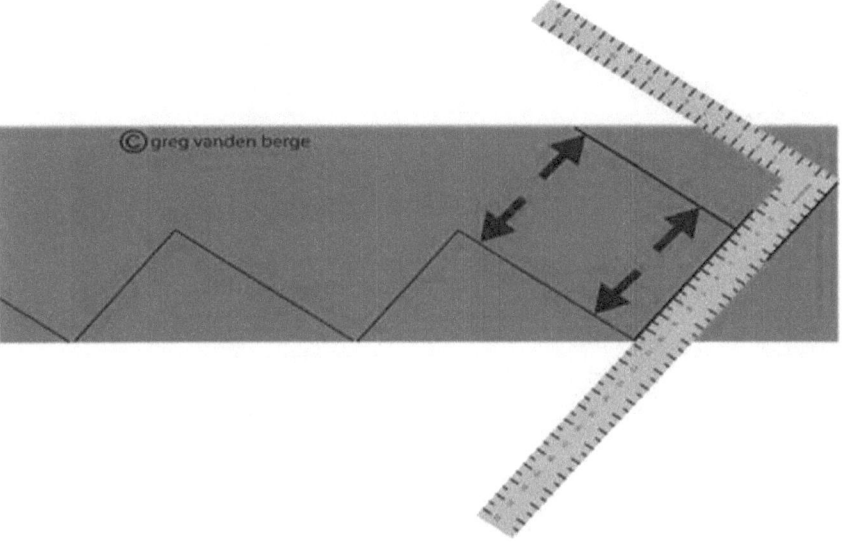

Step 2: Then we will mark the lumber using the front of the framing square as shown in picture above.

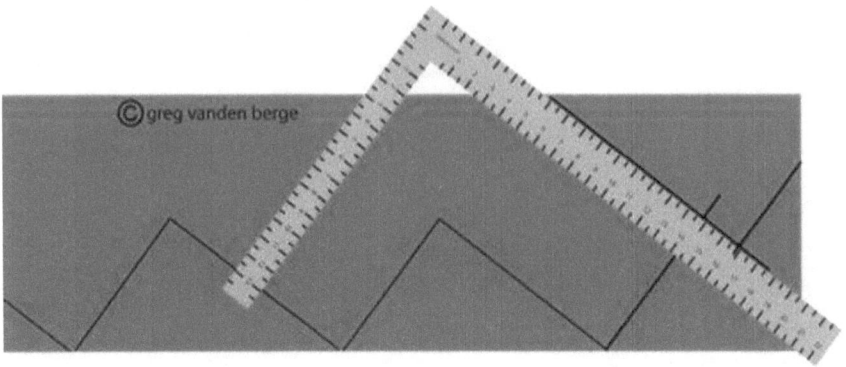

Step 3: In the next step we will extend the bottom line as shown in picture above. This line will represent the top of the landing as long as the stair treads and landing sheathing will be the same thickness.

For more information on riser variations go to page 80.

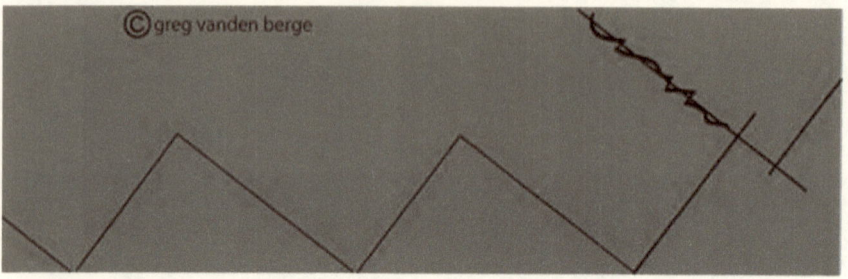

Step 4: Cross out any lines you aren't planning to cut.

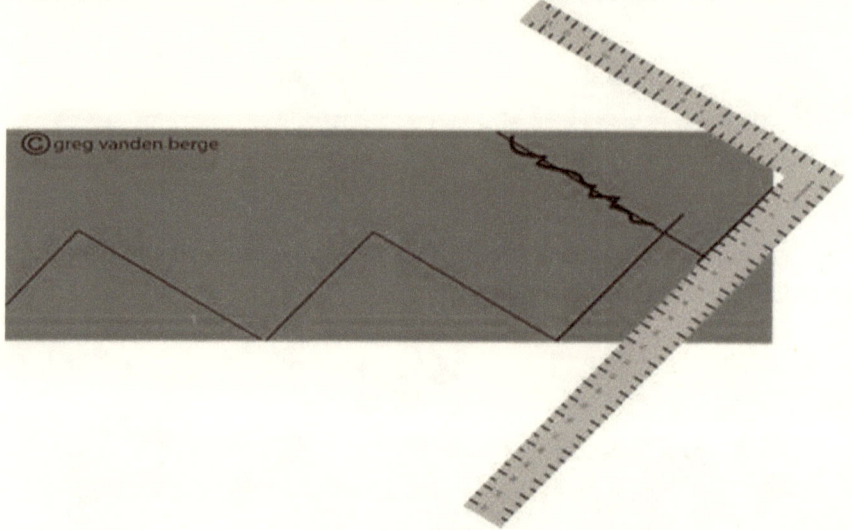

Step 5: The next step is critical for this particular layout. The measurement you're looking for will represent the thickness of the joist you'll be using to construct your landing.

In our example we'll be using 2 x 6 with a height of 5 1/2 inches.

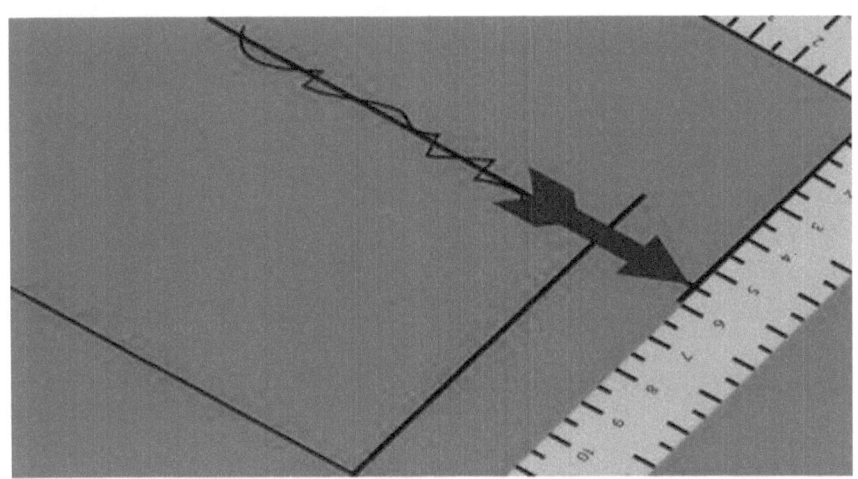

Step 6: Then you'll need to line the mark representing the top of joist up with the 5 1/2 inch measurement mark on the framing square.

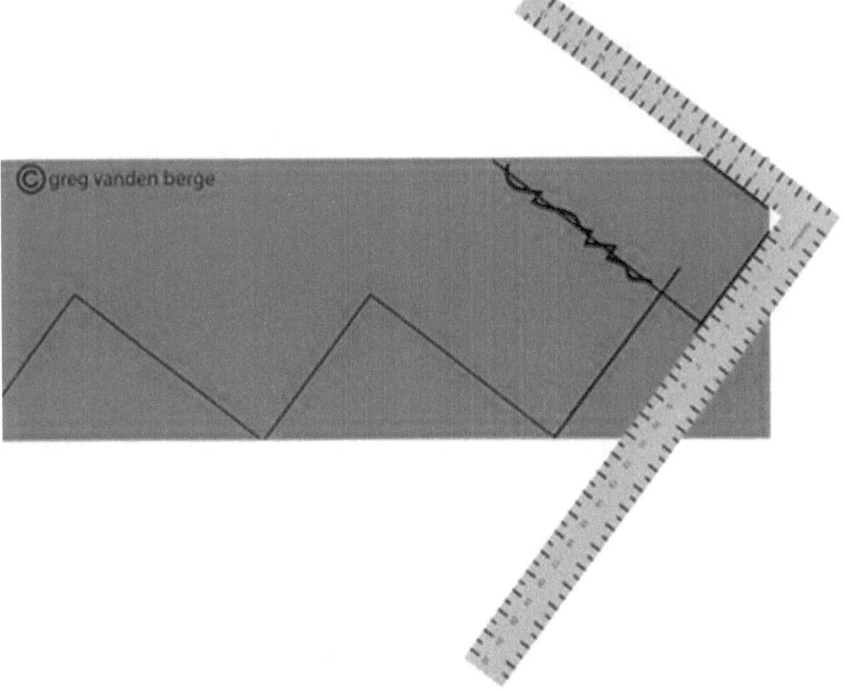

Step 7: After you've position the framing square correctly, mark the stringer bottom.

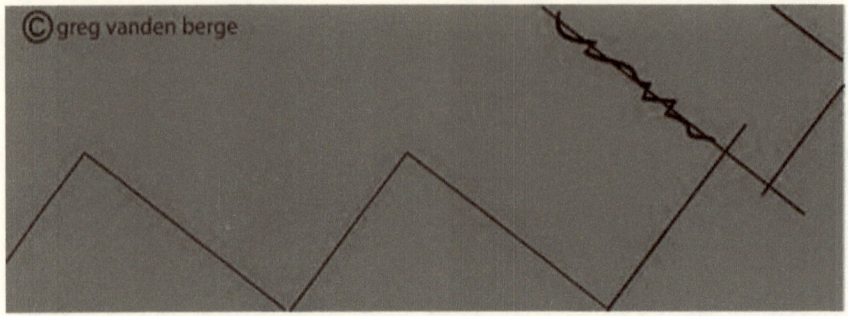

Step 8: You should end up with something like this.

Step 9: This is what the stringer will look like after it was cut. In the next steps I will provide you with a little more insight on why you did what you did.

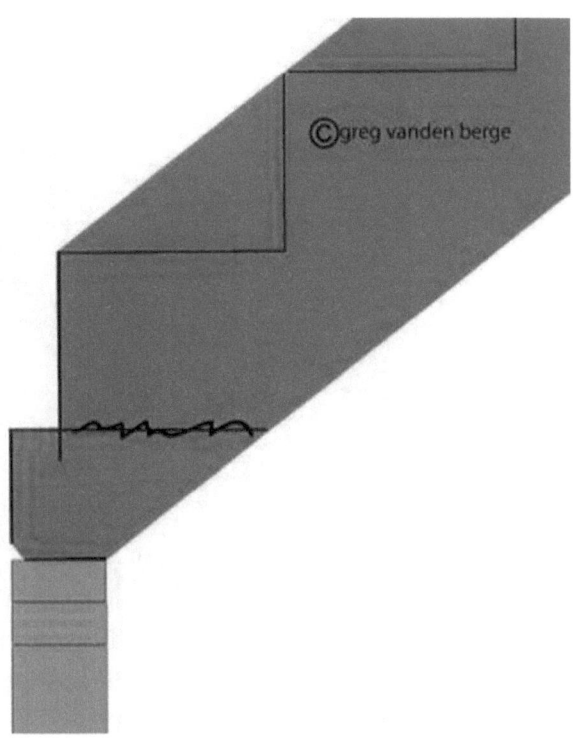

Step 10: In the illustration above the stair stringer bottom is sitting directly on top of a wall you would have framed to the correct height.

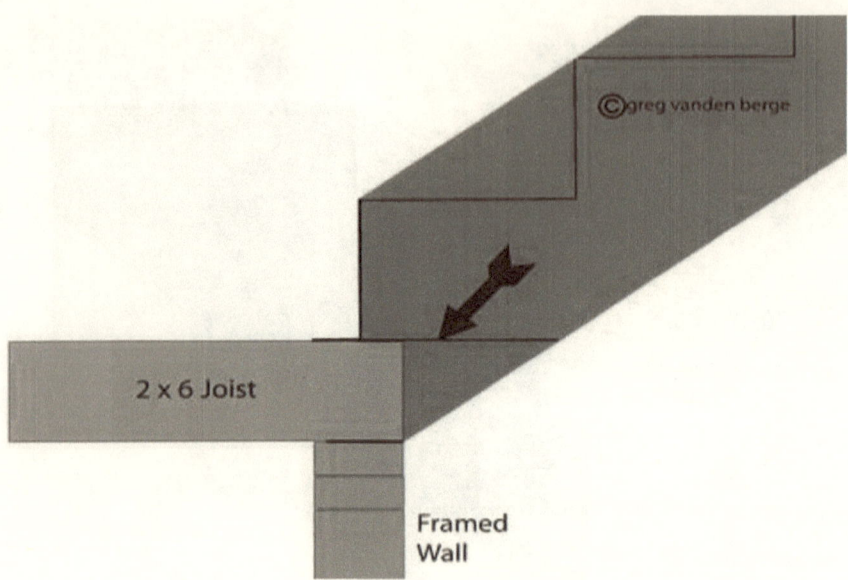

Step 11: As you can clearly see in the illustration above, the line representing the top of the landing joist, lines up perfectly with the top of the framed landing. Sometimes a picture can be worth a million words and if you didn't understand what you were doing earlier, when laying out your stringer, hopefully by now you do.

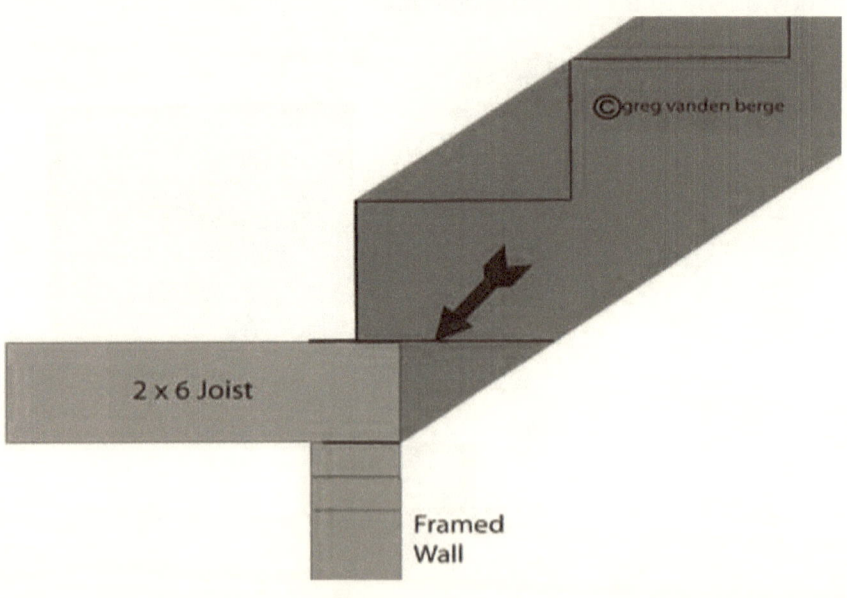

Step 12: In step number 11 the joist was placed in front of the stair stringer, but in this step it's placed behind the stringer to give you a better idea, what the finished framing will look like.

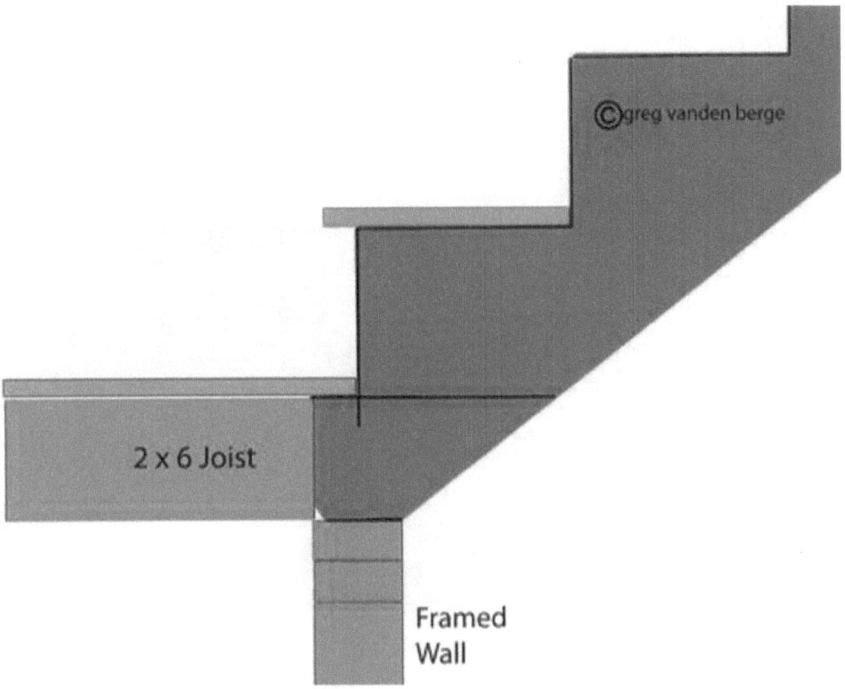

Step 13: This illustration provides you with a side view of the stairway after the landing was covered with a 3/4" thick piece of plywood and the stringer with a 3/4" thick stair tread.

On Top Of Landing

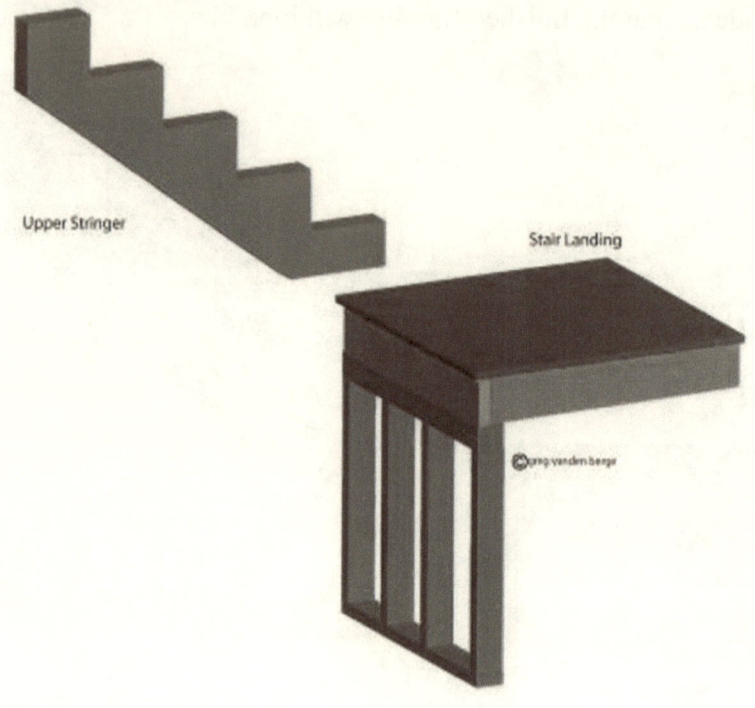

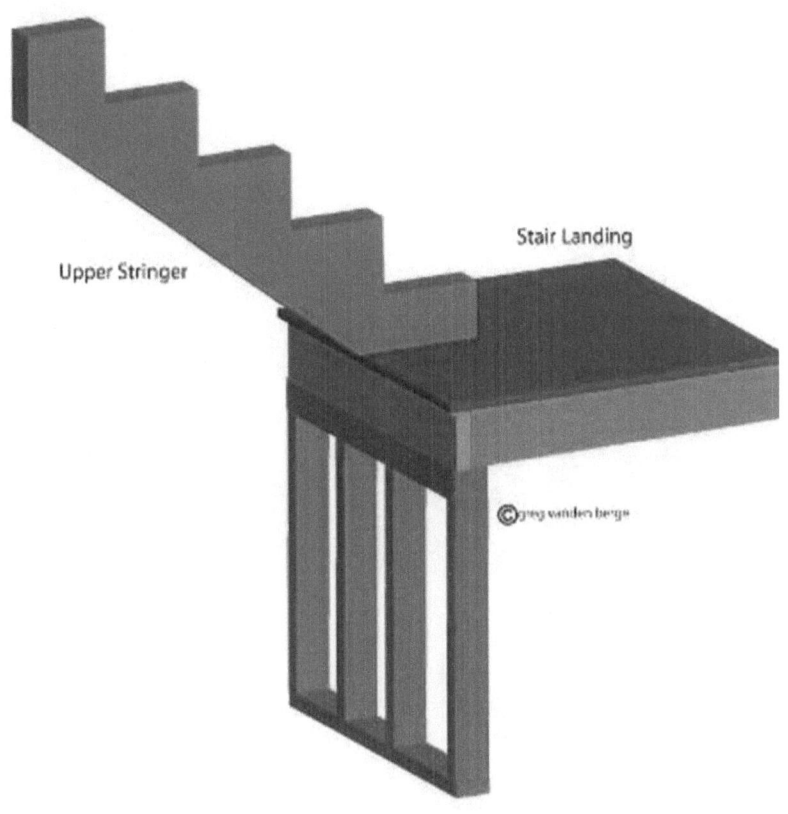

For Layout Instructions Follow <u>Steps 13 -16 and 18</u>

Hanger Method

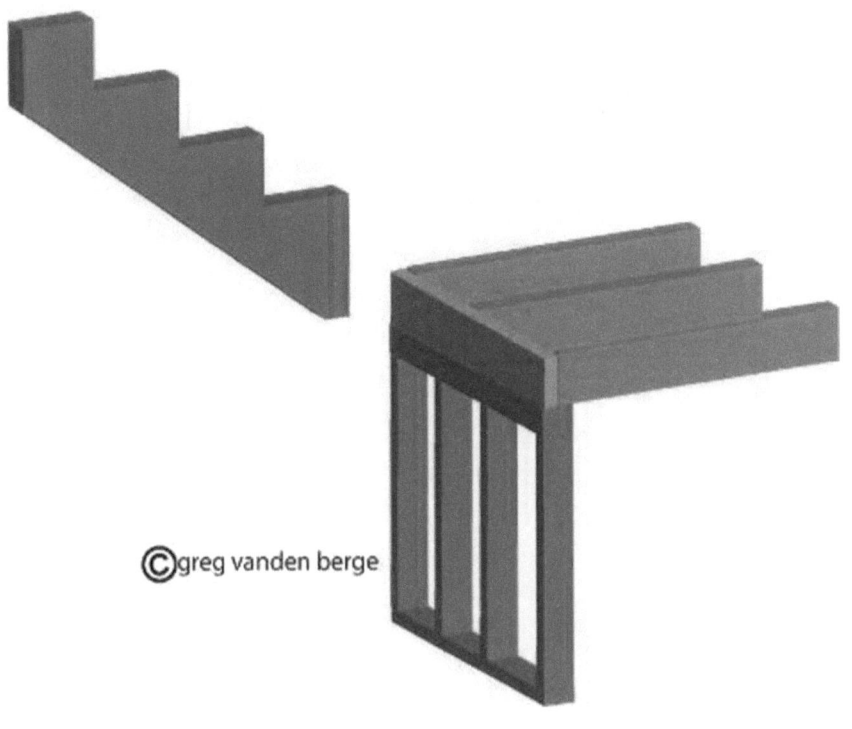

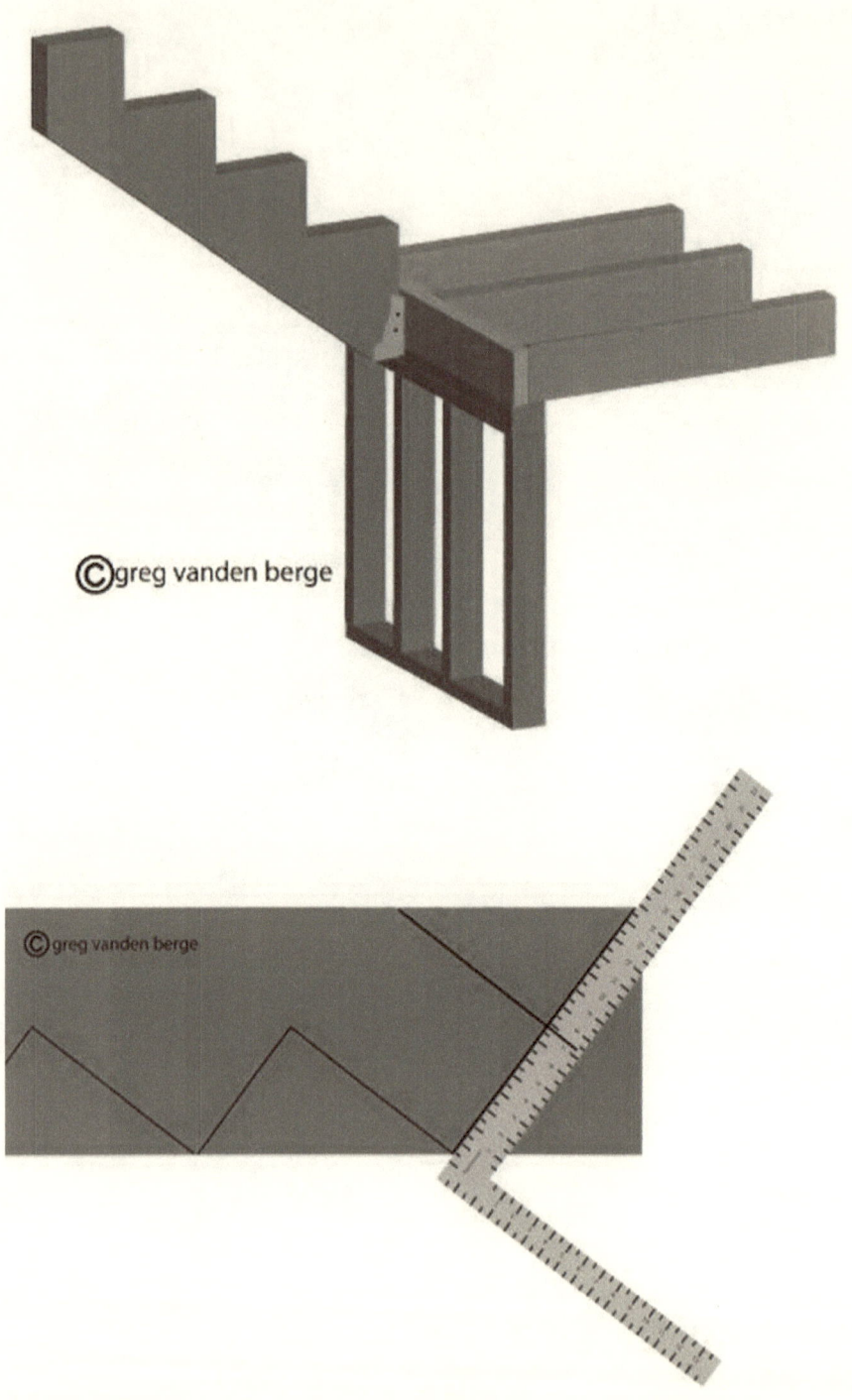

Step 1: Line framing square up with first riser as shown in picture above and mark.

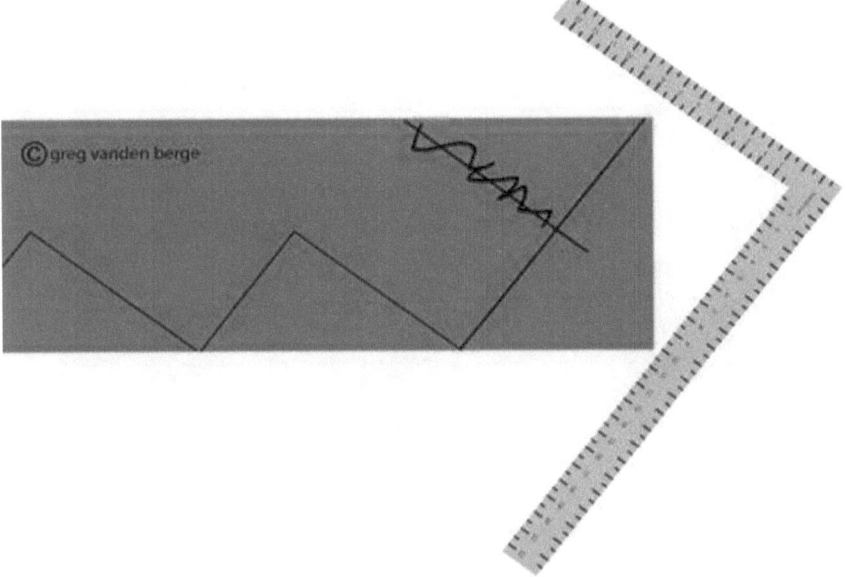

Step 2: Then cross out adjusted riser line as shown in picture above. Refer back to step 16 - 18 for more information on where to position adjusted riser line.

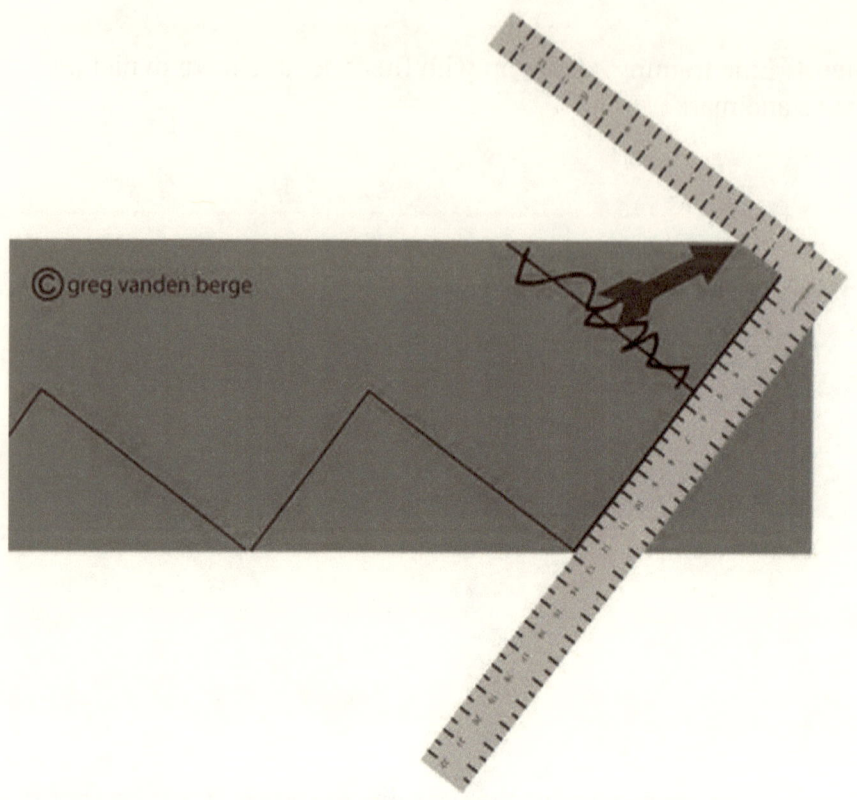

Step 3: In the next step you'll need to measure the distance of the bottom hanger you're planning on using.

If you're going to use a standard 2 x 10 joist hanger, then simply measure the distance the bottom of the stair stringer will be sitting in the joist hanger, when positioned against the landing.

The black arrow is pointing to the 2 inch mark on the framing square. If the distance is greater or less simply make the necessary adjustments, then position the framing square accordingly and mark.

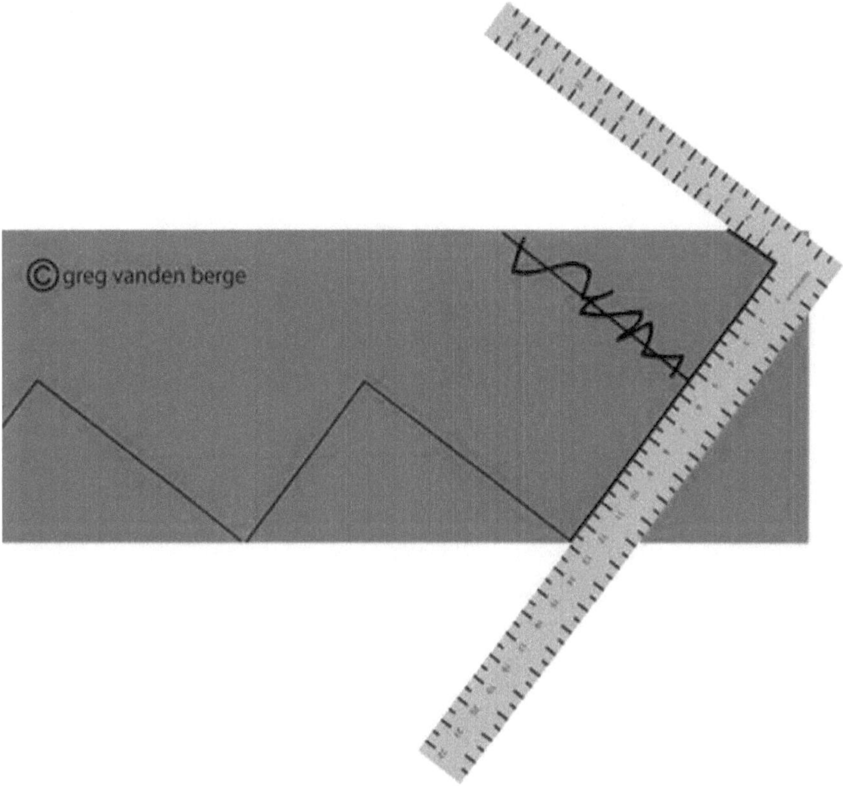

Step 4: You should end up with something like this.

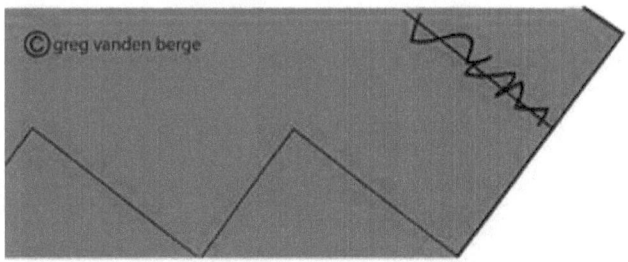

Step 5: I went ahead and cut the bottom of the stair stringer and will be properly positioning the stringer in the next steps, to provide you with a better understanding of why you did what you did.

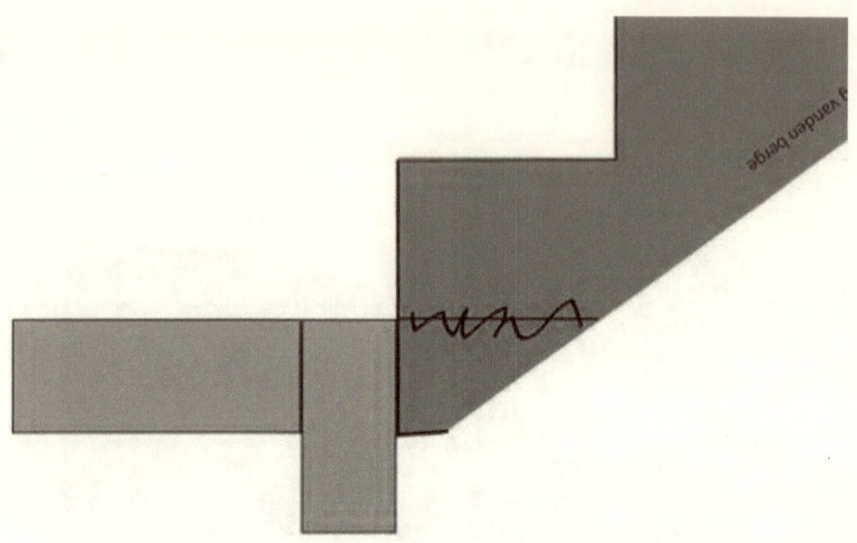

Step 6: This is what the stringer would look like when it's located in its proper position. If you notice the adjusted riser line, lines up with the top of the landing or floor.

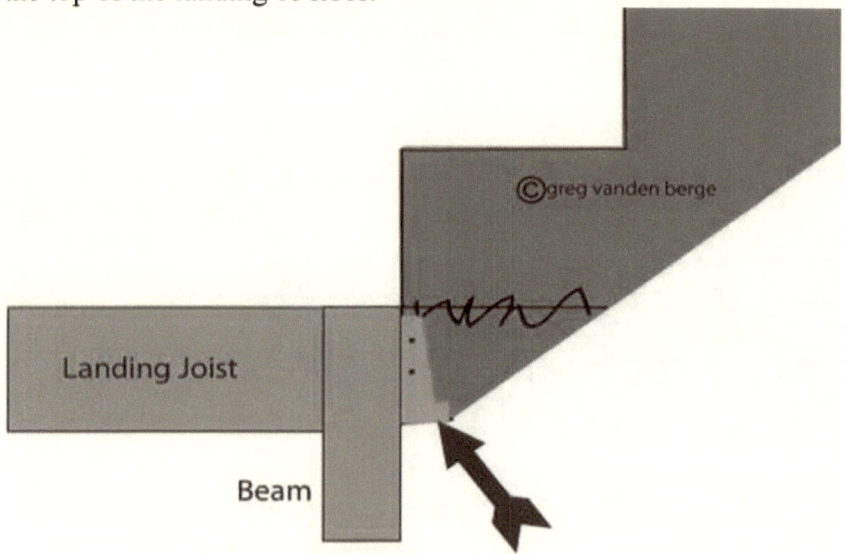

Step 7: The illustration above provides you with an example of the stairway, after the hanger and stringer have been positioned properly.

Top Stringer Layout Methods

Beam And Ledger

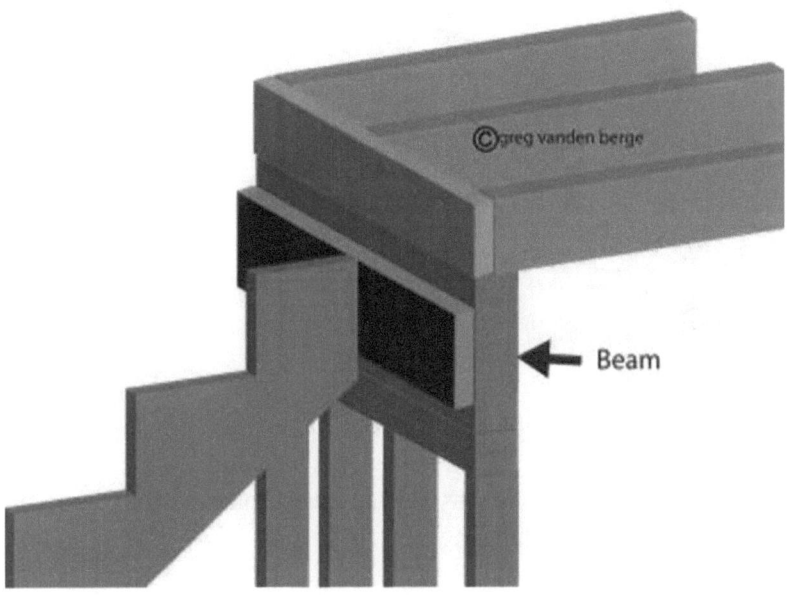

For Layout Instructions Follow Steps 19 -24

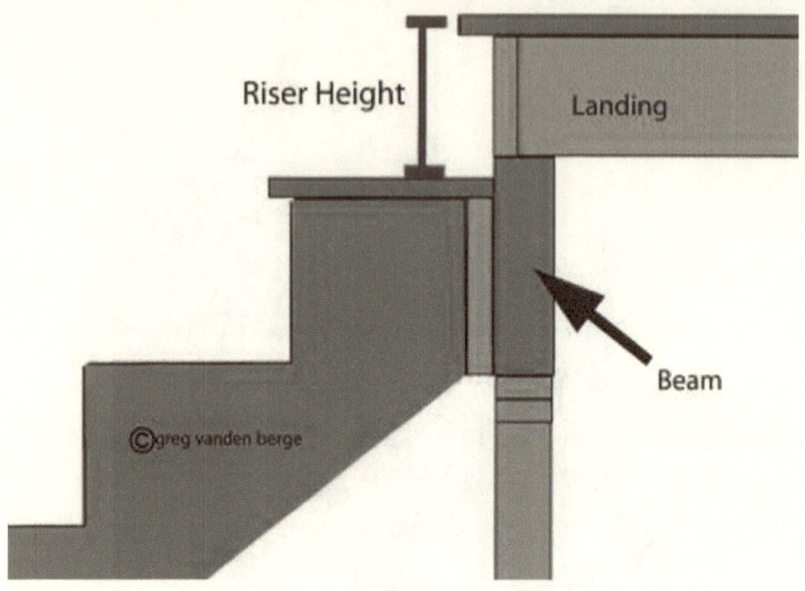

1. This is a side view of a stairway with a beam that's sitting directly on top of a wall, under the landing.

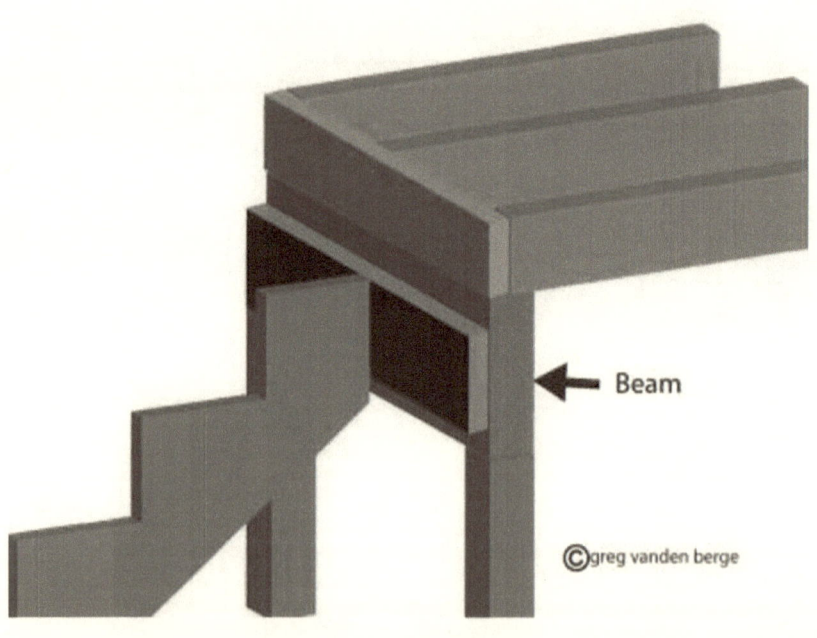

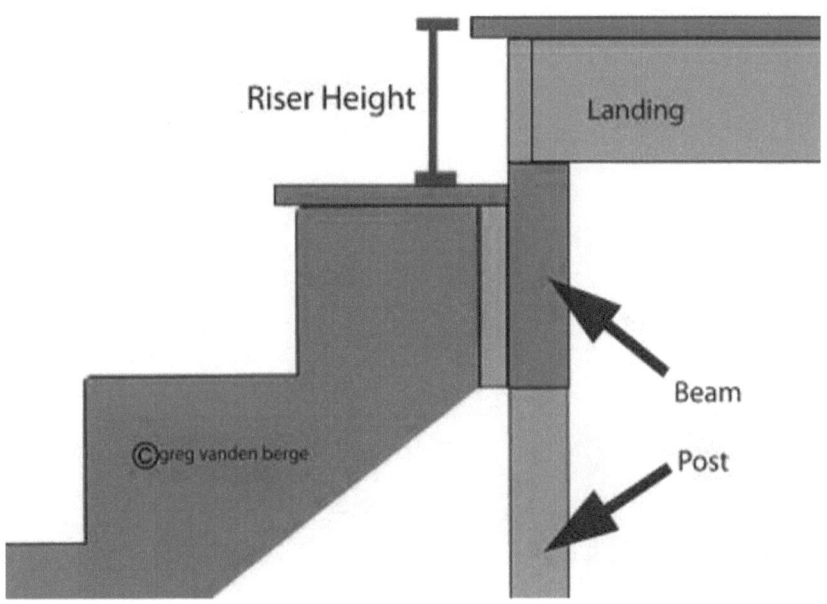

2. This is a side view of a stairway with a beam that's sitting directly on top of two posts, under the landing.

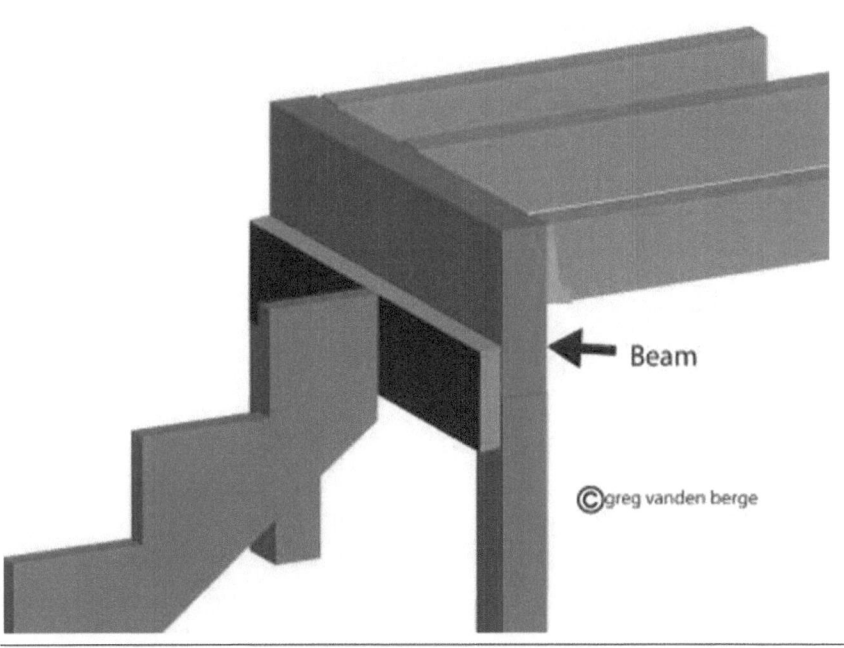

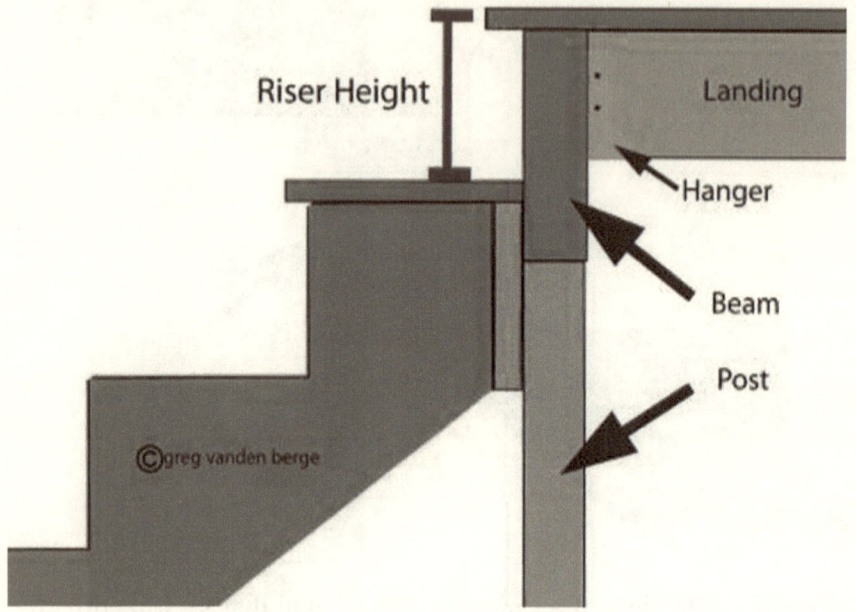

3. The picture above provides us with a side view of a stairway, with a beam that's sitting directly on top of a post, but built into the landing.

I provided you with these three illustrations to give you a few more options, while planning and designing your stairway and stair stringer layout.

Hanging Plywood Ledger

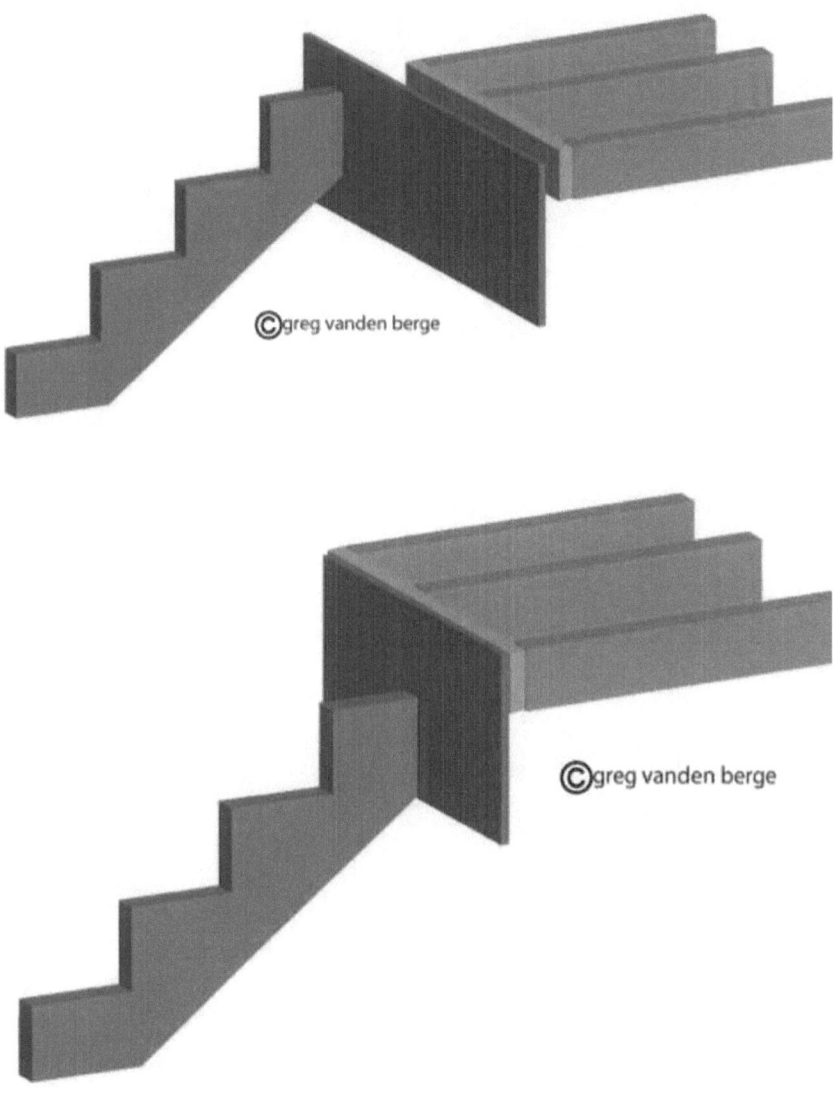

Even though I'm providing you with this method, I've never used it and probably never will. However, I've seen it used before, so here it is.

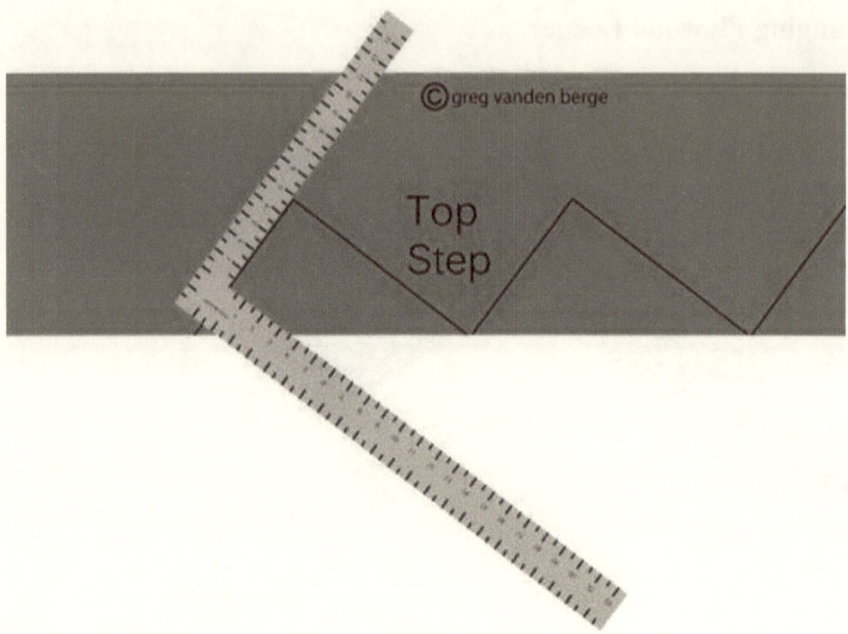

Step 1: Line the framing square up with the last riser as shown in picture above.

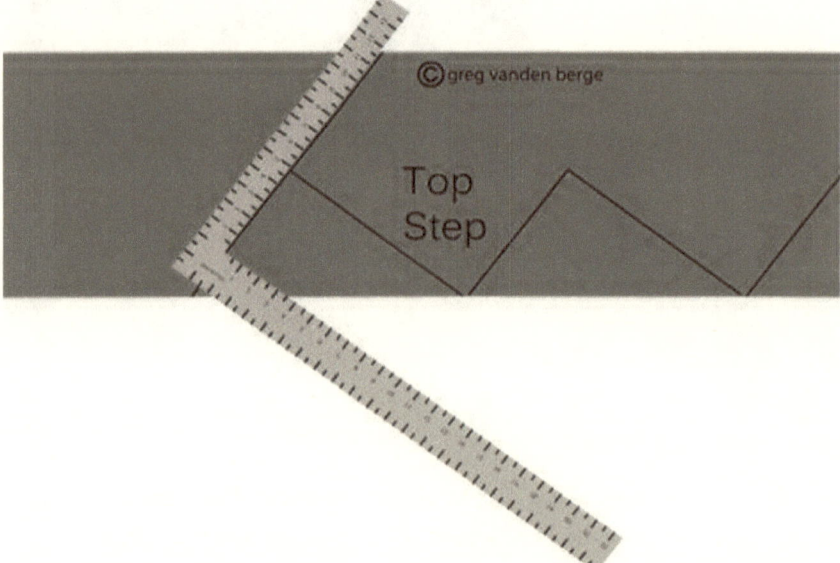

Step 2: When the framing square is in the correct position, mark lumber.

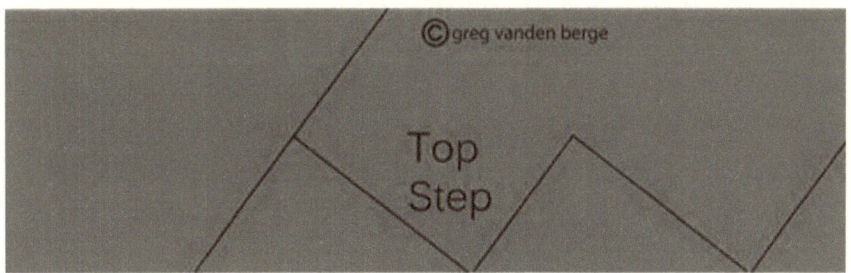

Step 3: You should end up with something like this when you're done.

Step 4: For this method you won't need to deduct materials for a ledger. The illustration above provides you with an excellent example of what your stringer would look like after being cut.

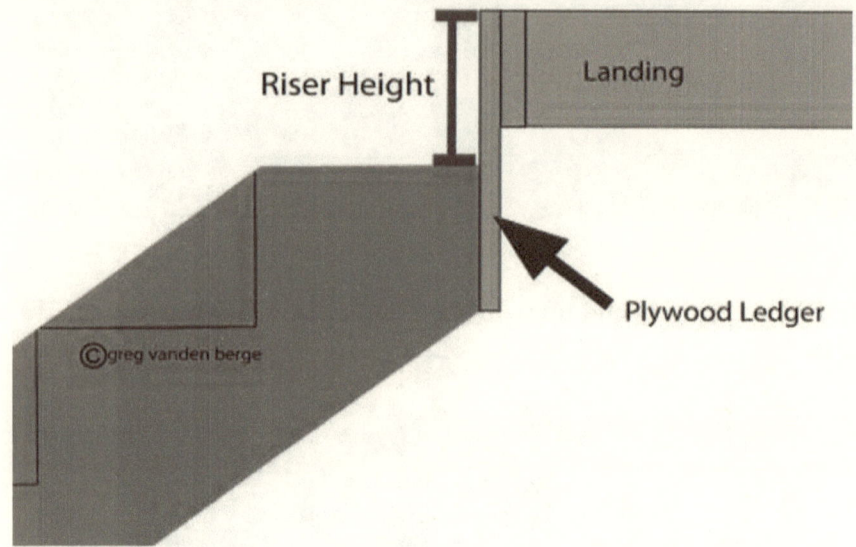

Step 5: Here's a side view of the hanging ledger, landing and stair stringer.

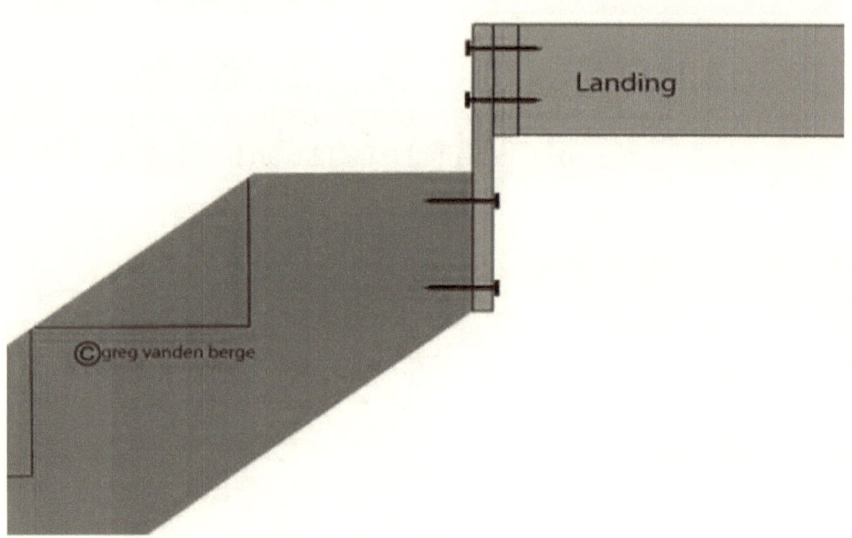

Step 6: The illustration above provides you with a method for nailing and attaching your plywood ledger to the stair stringer and landing.

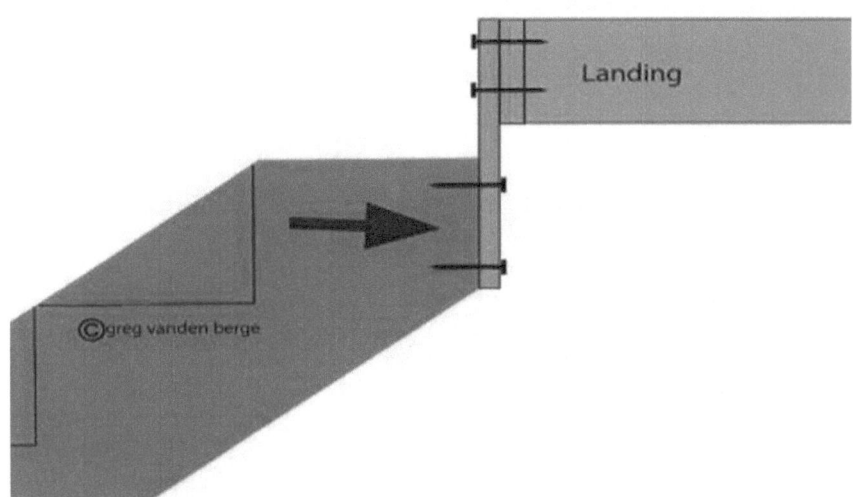

Step 7: Here's one of the reasons why I don't use plywood ledger's for situations like this. Under the right conditions stairways like these will move and over time this movement could create problems.

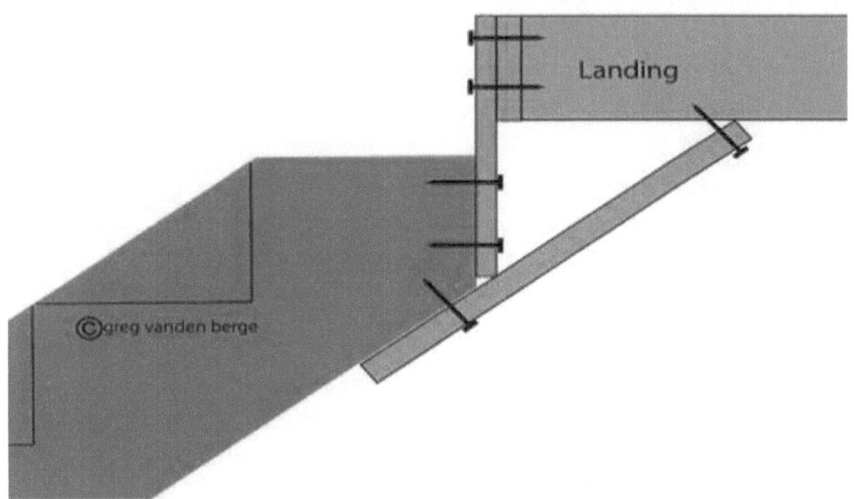

Step 8: If you are planning on using this method then it wouldn't be a bad idea to brace your stairway with additional supports as shown in picture above. Simply nail a 2 foot board to the bottom of the stair stringer and landing.

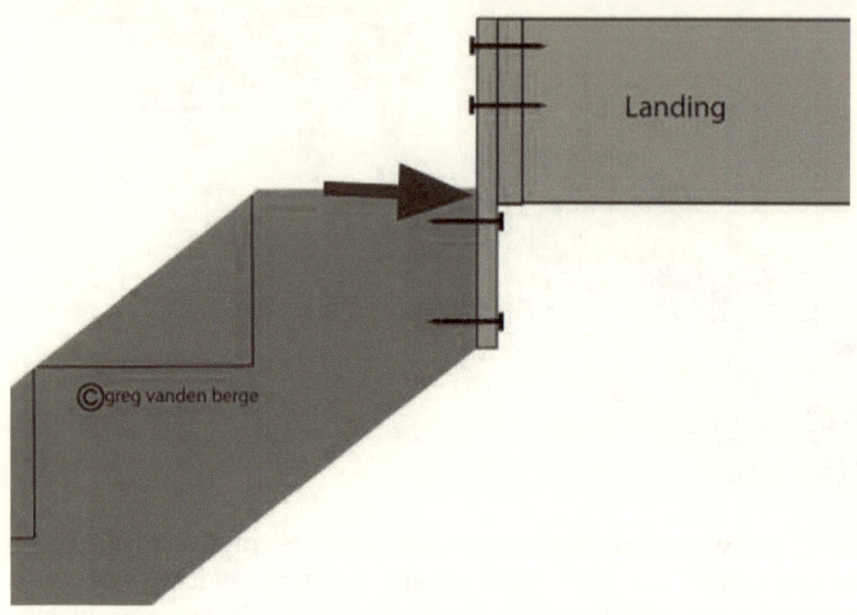

If your landing joist are a little longer than your riser, as shown in the picture above then bracing wouldn't be necessary. Even though it isn't much, something like this would be enough to prevent horizontal movement.

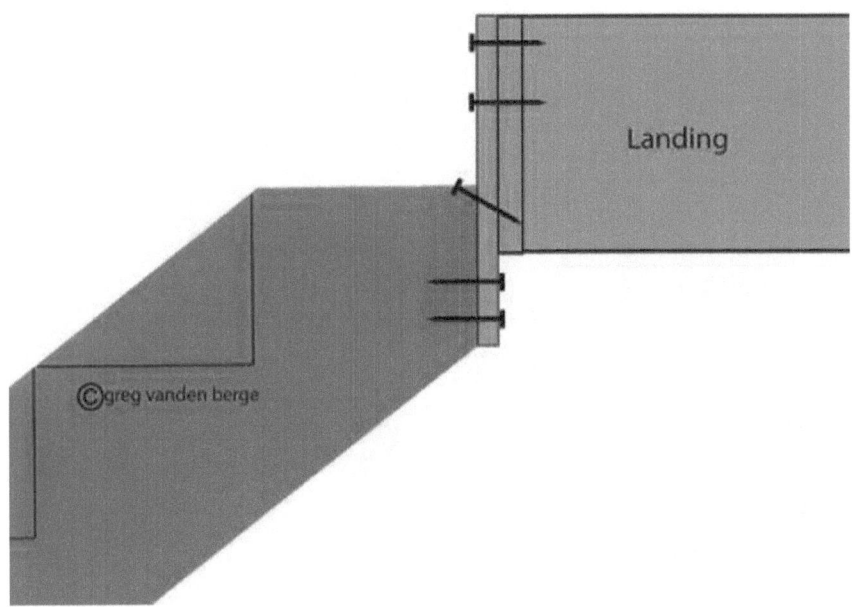

Larger joist provides us with more lateral structural support, but creates other problems. As you can see in the picture above, there isn't a lot of room to nail the back of the plywood ledger to our stringers.

If the joists are too big, then you would need to nail the plywood to each stair stringer before nailing the ledger to the landing.

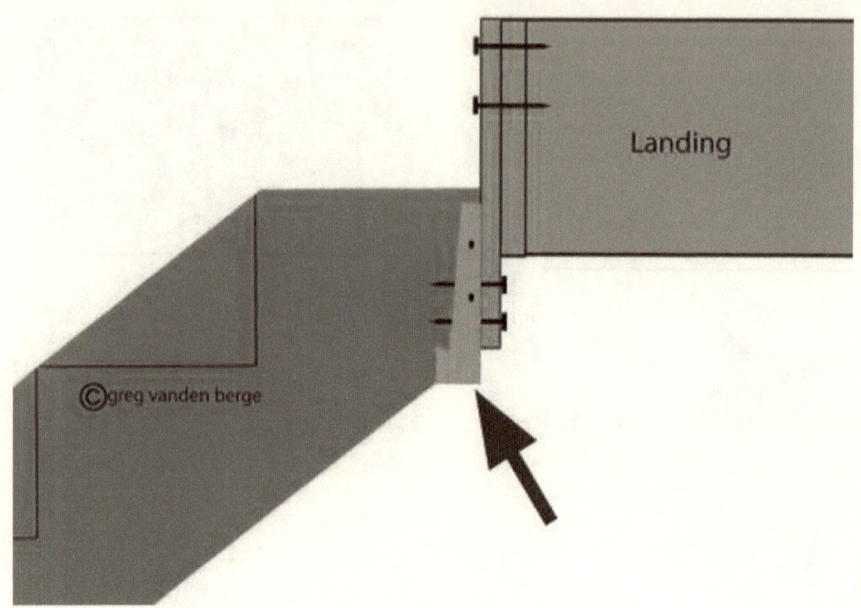

For additional structural support you could always use a metal hanger.

Attaching Stringer To Joist

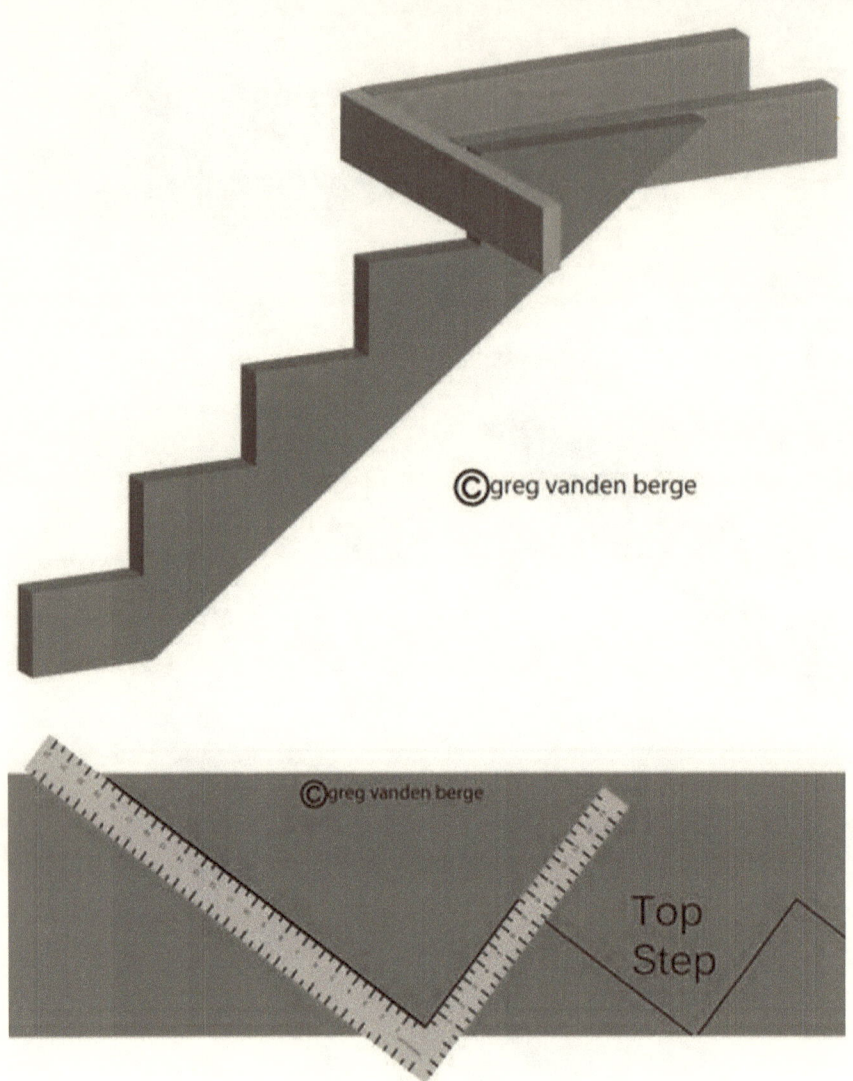

Step 1: Line the framing square up with top riser as shown in picture above.

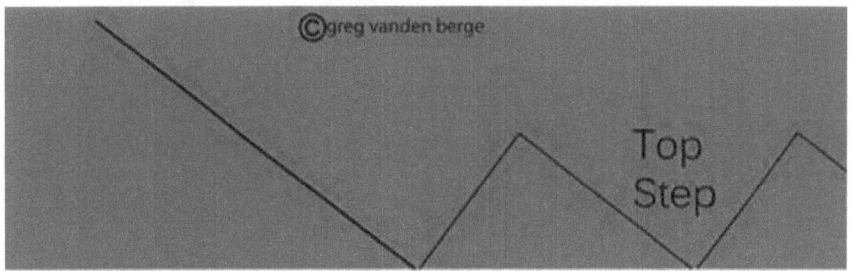

Step 2: You should end up with something like this.

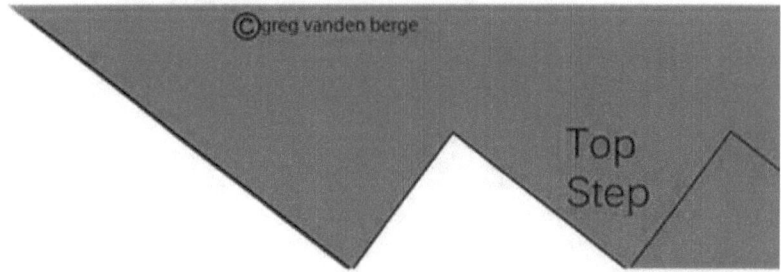

Step 3: In this illustration I cut the stringer and will be rotating it into its proper position to give you a better idea through the next steps, why you needed to do what you did.

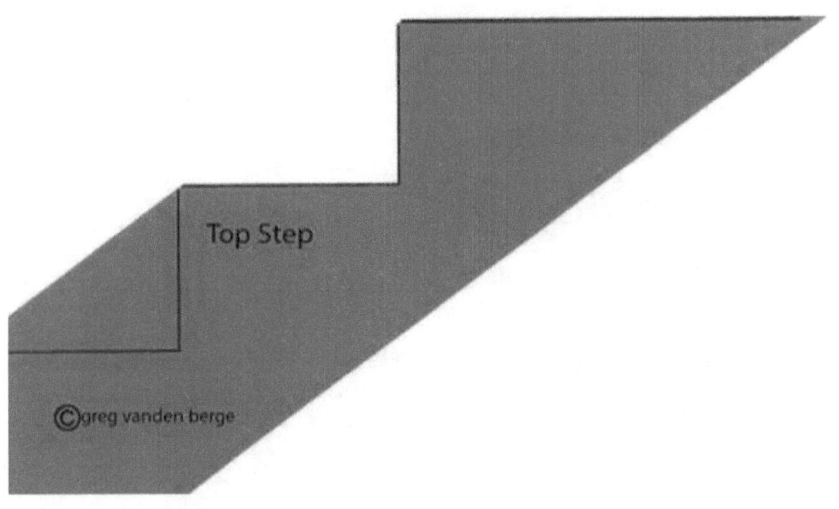

Step 4: In this illustration the stringer has been turned over into its proper or upright position. Again, some of these illustrations will put all of the pieces to the puzzle together and help you make sense out of the nonsense.

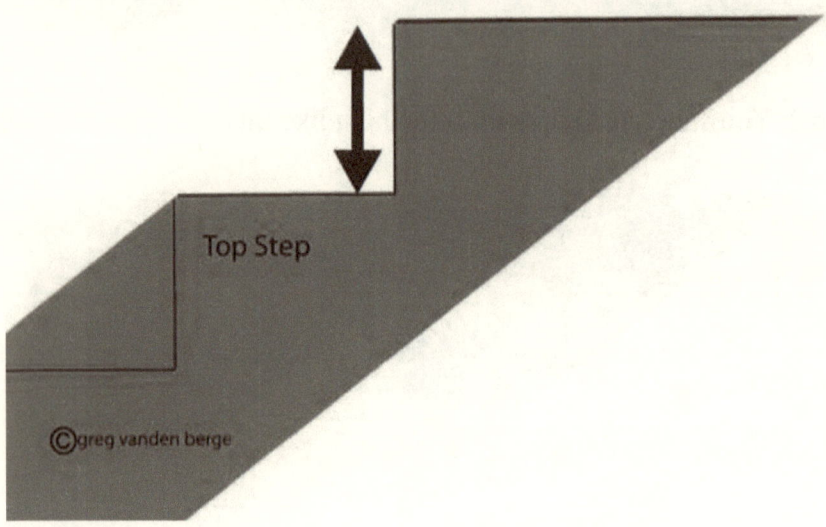

Step 5: Don't forget to adjust the riser layout if the landing or deck sheathing will be a different thickness than your stair treads.

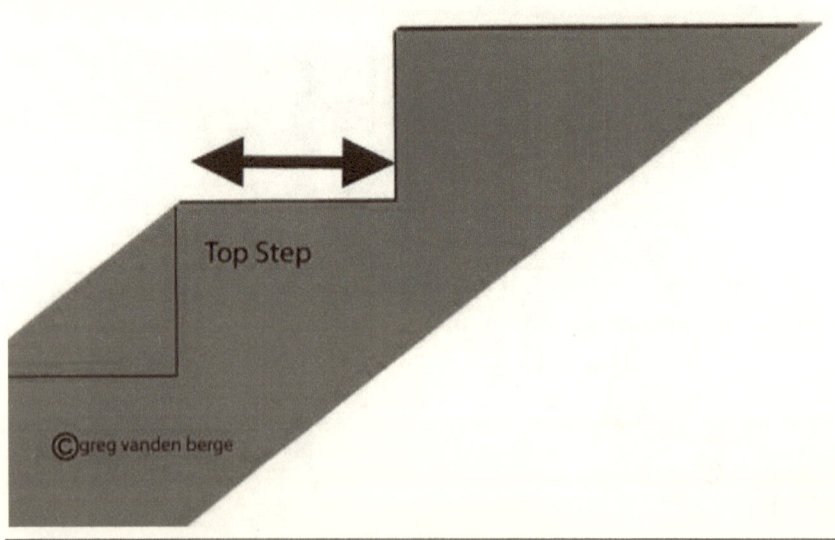

Step 6: If you're going to use different sized materials you will need to compensate for them when laying out your stair stringer.

For more information on riser variations go to page 80.

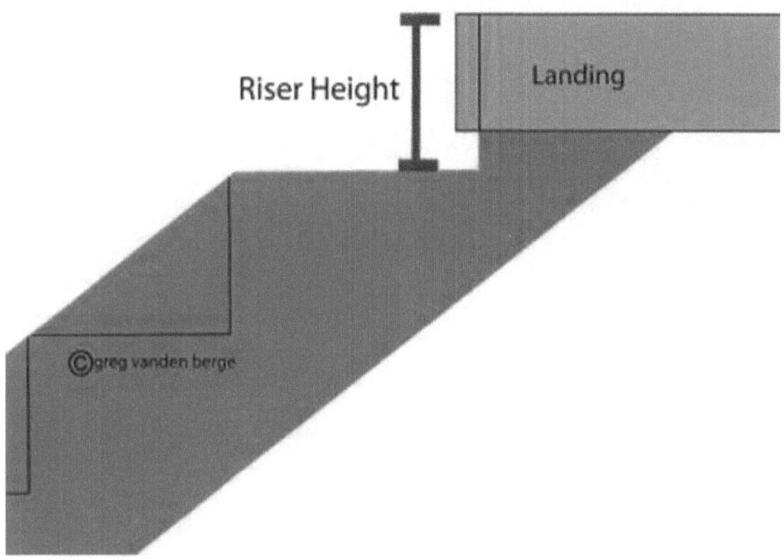

This is what the stair stringer would look like with one of the landing joist on the front side.

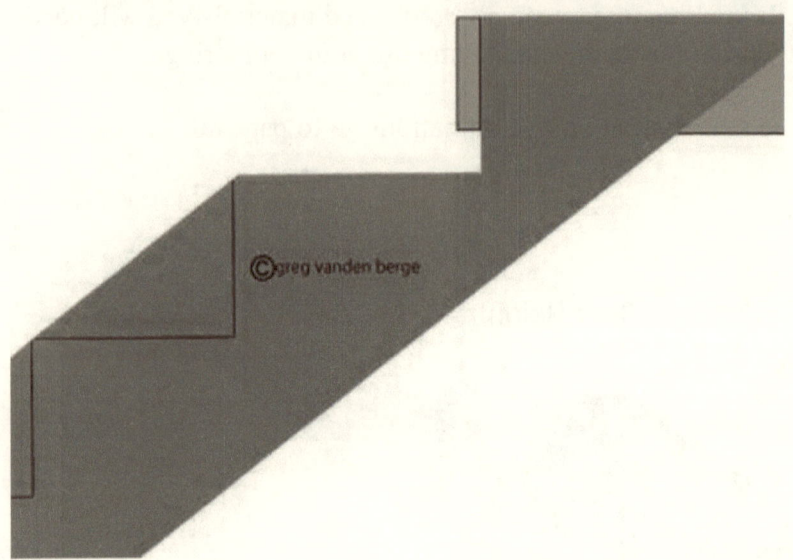

Here's what it will look like on the other side. This type of stair stringer design can simply be nailed to the landing or decking joist.

This application might not work under certain circumstances. However if you were building the stairs and landing at the same time, the joist could be laid out to accommodate the stair stringer's.

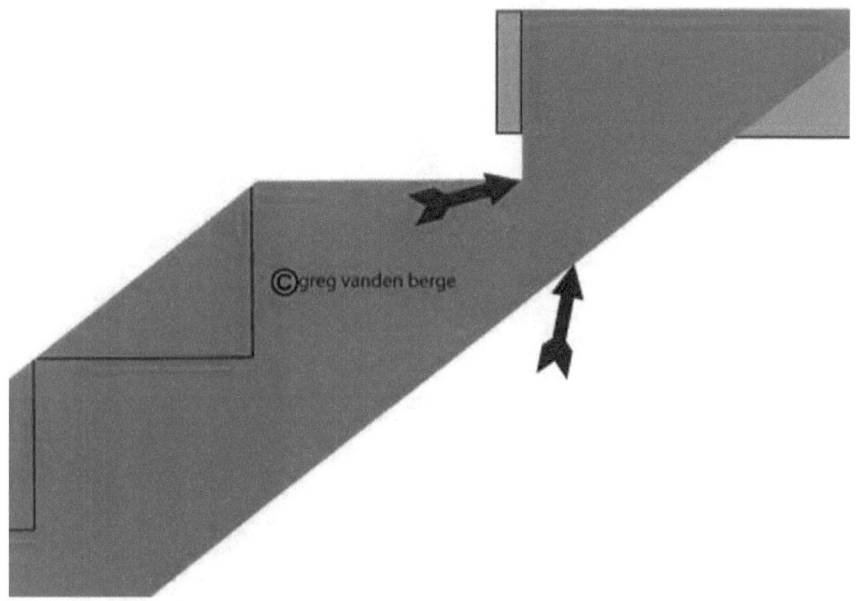

Here's another problem you should look for when working with this design. The distance between the arrows shouldn't be less than 5 inches.

The situation in the illustration above could become a problem when working with certain types of stairways, decks and landings. If your stairway or deck design creates any structural problems that will weaken the stairway in any way, then try to figure out a different way to layout the stringers and build the stairs.

In Floor Or Landing

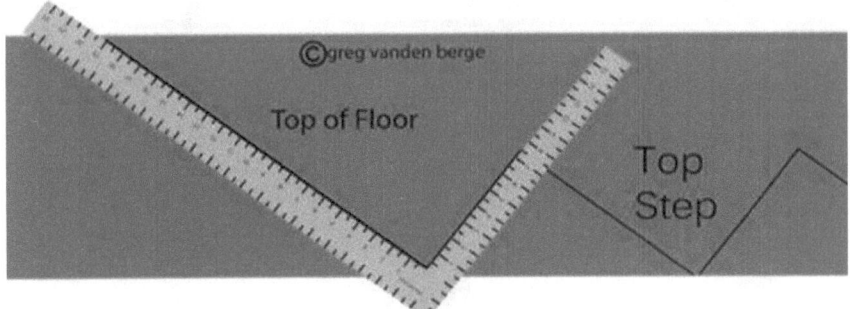

Step 1: Position framing square on the top riser as shown in picture above. Make sure the last riser height has been adjusted for any lumber thickness variations and then mark.

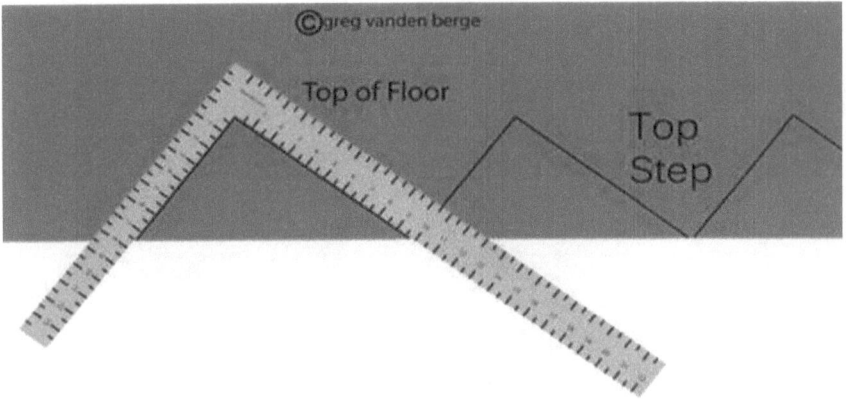

If you have enough room, while you're marking and laying out your risers and stair stringer, then you should always lay out an additional step.

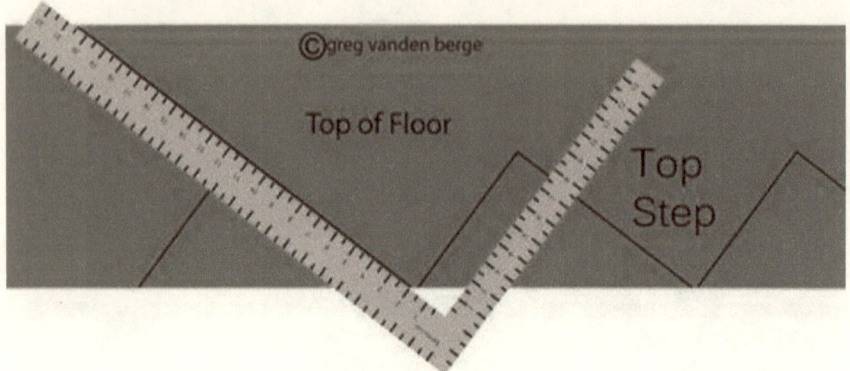

Then you can line the framing square up with the stair tread layout mark that will represent the floor or landing. Either one of these methods will work, but I preferred using this one.

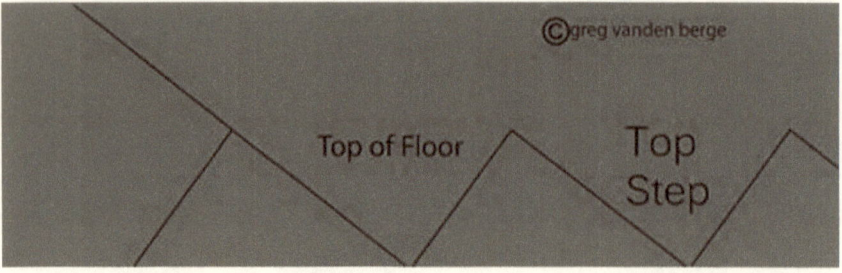

Step 2: After you've completed step one you should end up with something like this.

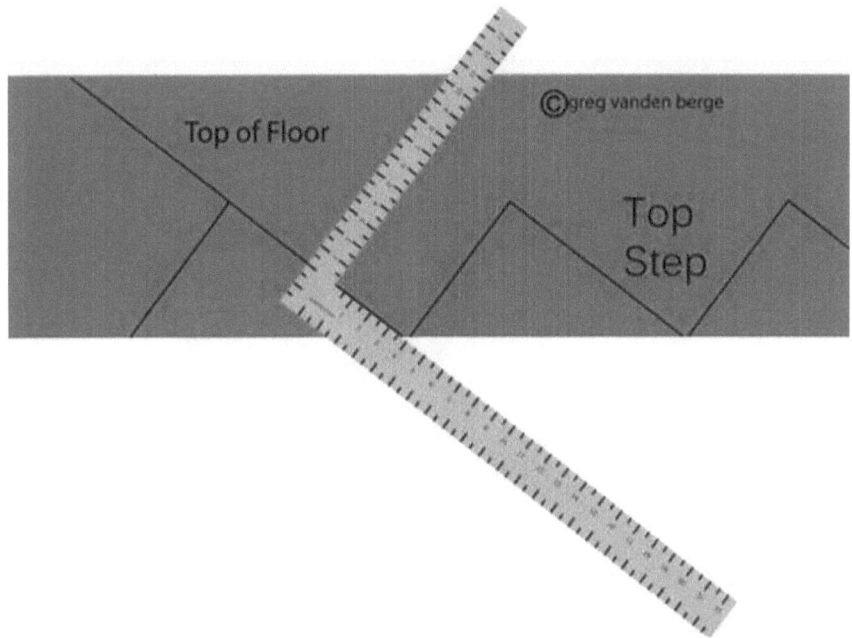

Step 3: In the next step you'll need to position the framing square, according to the width of the floor or landing joist. In the example above I'm lining up the 11 1/2 inch mark at the end of the framing square, up with the bottom or backside of the stair stringer.

In the next examples I will position the framing square in different areas, to provide you with a better idea of how to layout different floor joist widths.

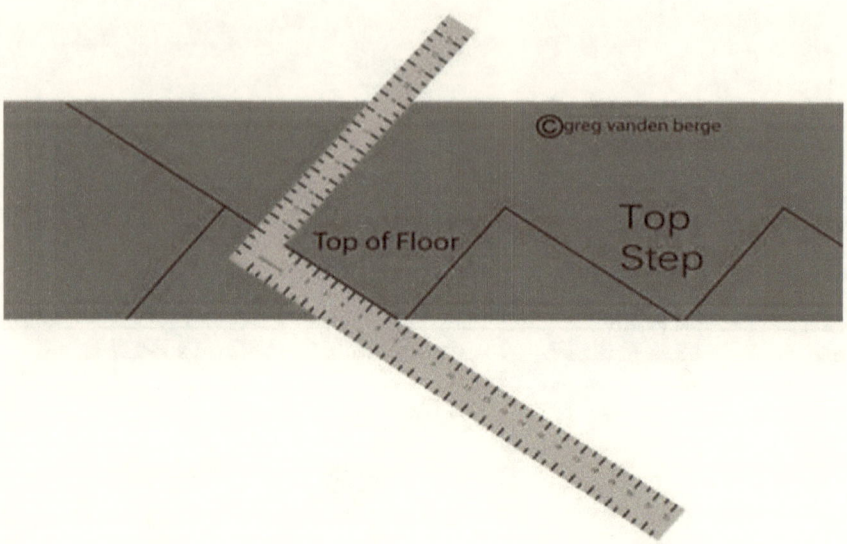

In this example I'm lining up the 9 1/2 inch measurement on the framing square, up with the back or bottom of the stair stringer. We would use this layout measurement for 2 x 10 floor or landing joist.

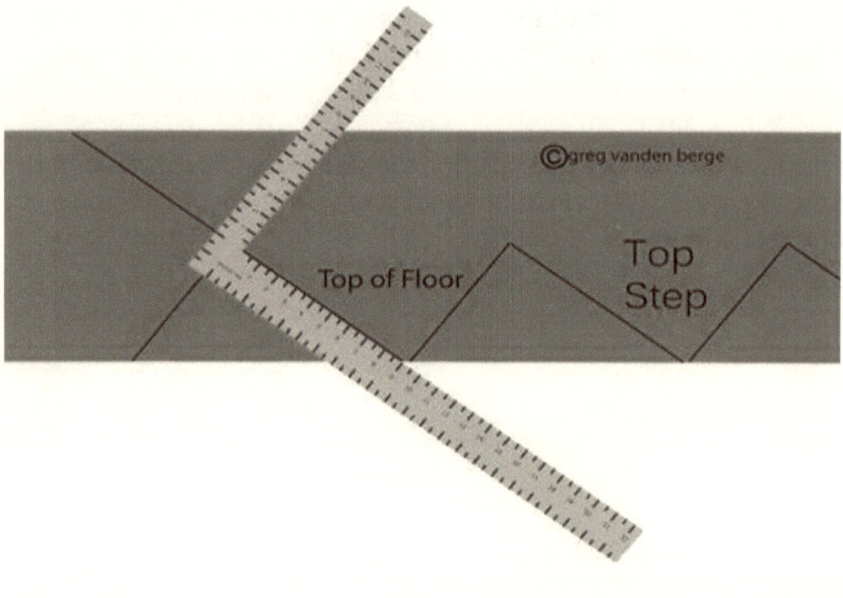

In this example I'm lining up the 7 1/2 inch measurement on the framing square, up with the back or bottom of the stair stringer. We would use this layout measurement for 2 x 8 floor or landing joist.

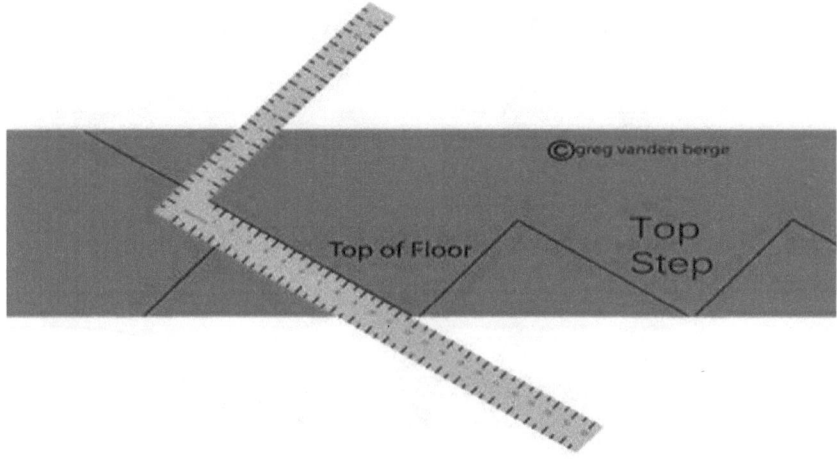

In this example I'm lining up the 5 1/2 inch measurement on the framing square, up with the back or bottom of the stair stringer. We would use this layout measurement for 2 x 6 floor or landing joist.

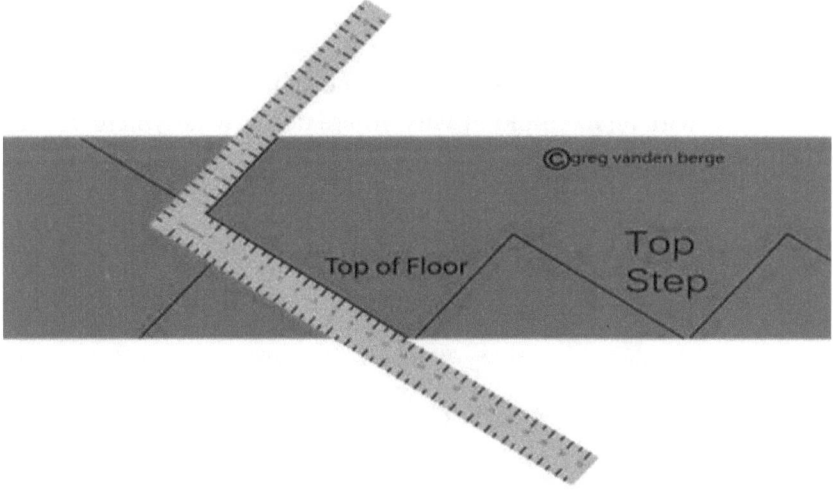

Step 4: In the next step we will select our joist measurement and mark the top of the stair stringer.

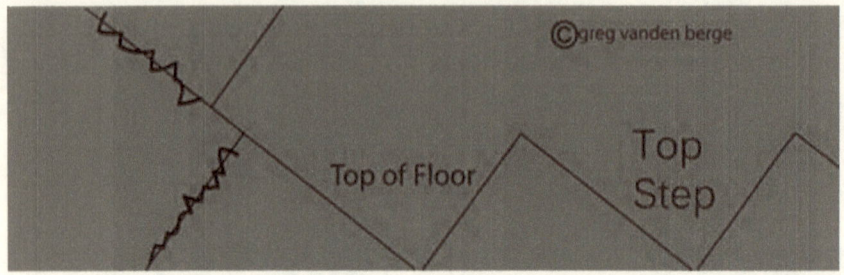

Step 5: Make sure you cross out any lines you won't be using or cutting.

Step 6: I cut the top of the stair stringer and will be positioning it, to provide you with an example of why you're doing what you're doing in the next illustrations.

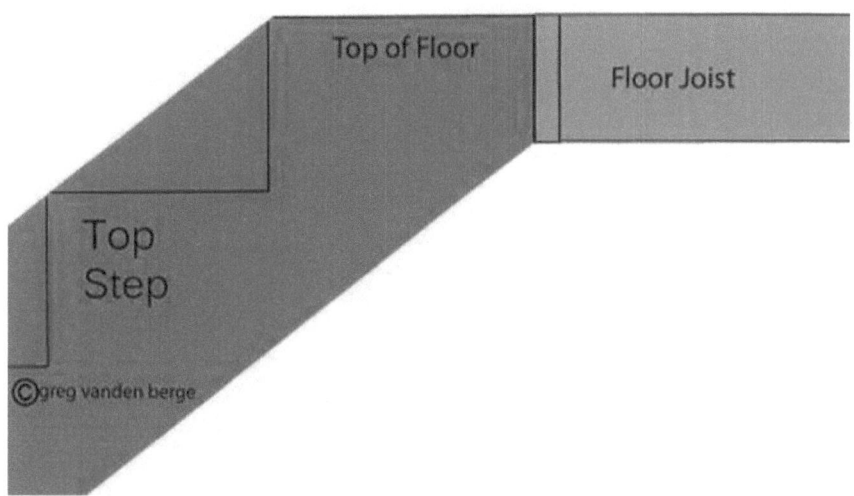

Step 7: As you can see in the illustration above, the mark we made representing the last tread or the top of floor, while laying out the stair stringer, lines up perfectly with the top of the floor joist.

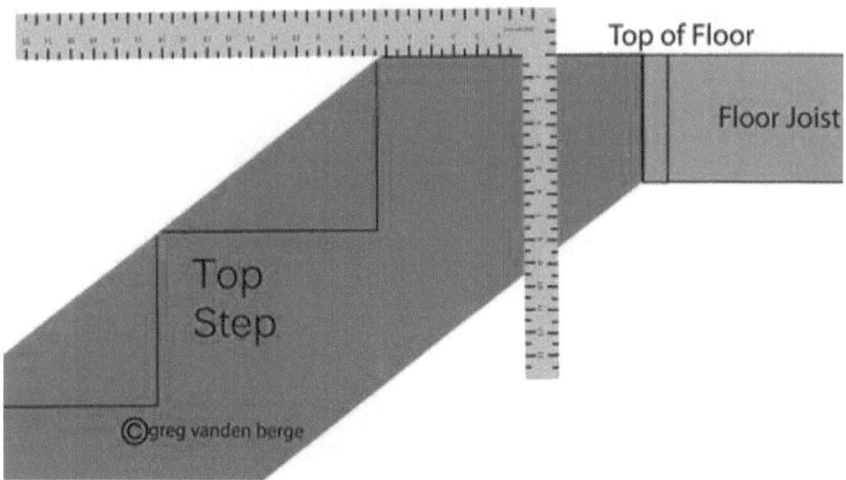

In this illustration I placed the framing square on top of the stair stringer, to provide you with a better idea of how we used 90° angles. One of the biggest problems with building stairs or laying out stringers is getting the proper perspective of why you're doing, the things you're doing.

Some of these illustrations can be viewed by some as unnecessary, while for others it paints a crystal-clear picture and will make all the difference in the world.

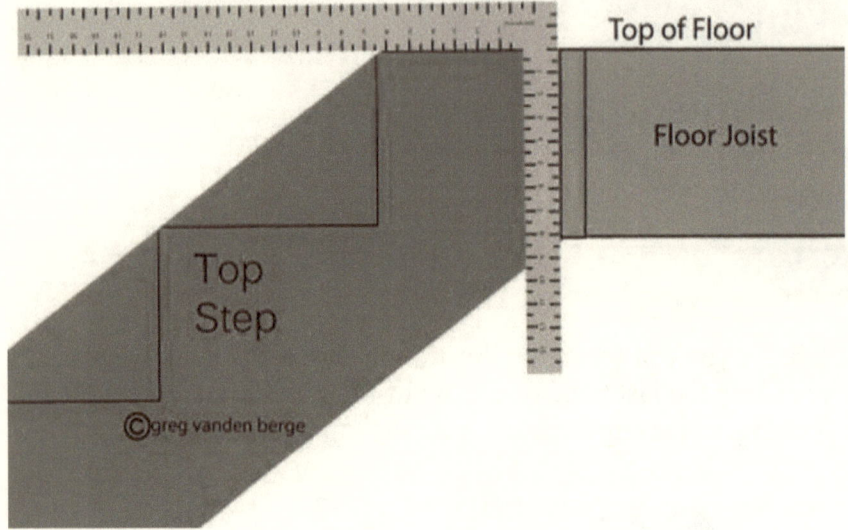

In the illustration above and below, I'm providing you with an example of what it would look like if the floor joist was a little wider.

For example, 2 x 10 or 2 x 12.

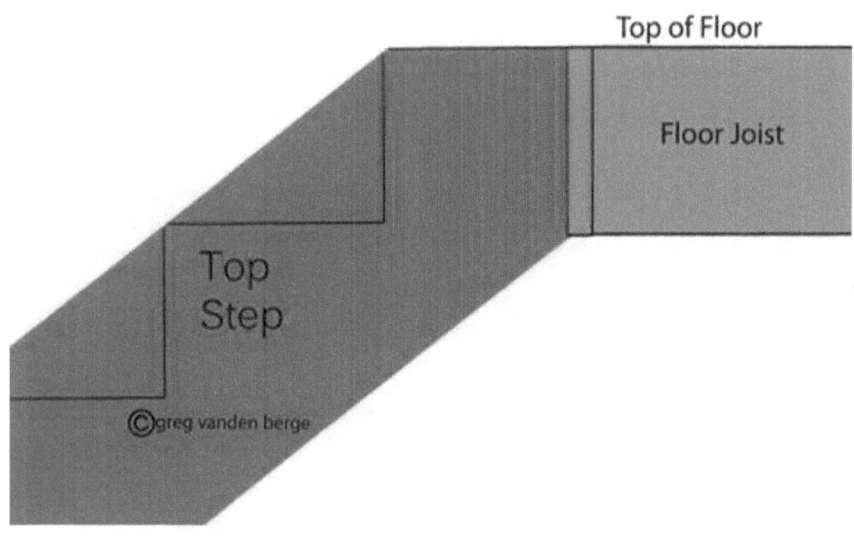

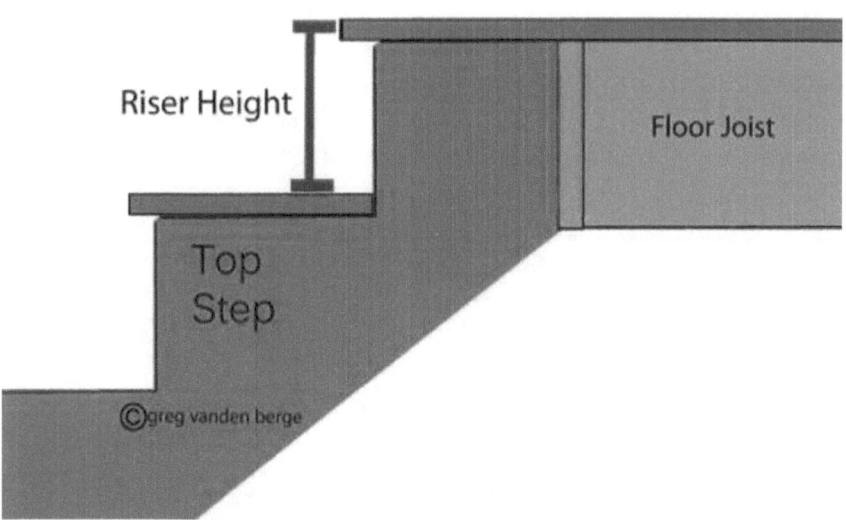

Don't forget to compensate for any variations in your stair tread or riser thickness. The main objective for laying out your stair stringer will be to lay them out, so each step is exactly the same size.

Layout Tips And Rules

90 Degree Angles

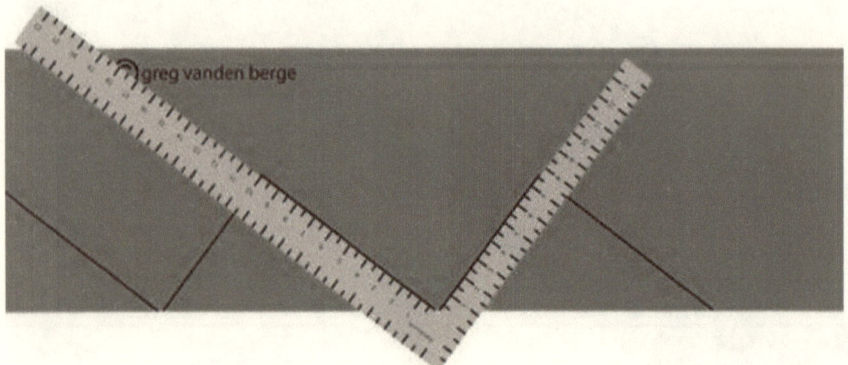

Practically every mark you make on your stringer will be at a 90° angle. The illustration above provides us with an excellent example of what I'm referring to.

Treads will represent horizontal marks and risers will be vertical, forming 90° angles. Understanding this concept can really make and I mean really make a big difference in whether or not your stairway is built correctly.

Parallel Lines

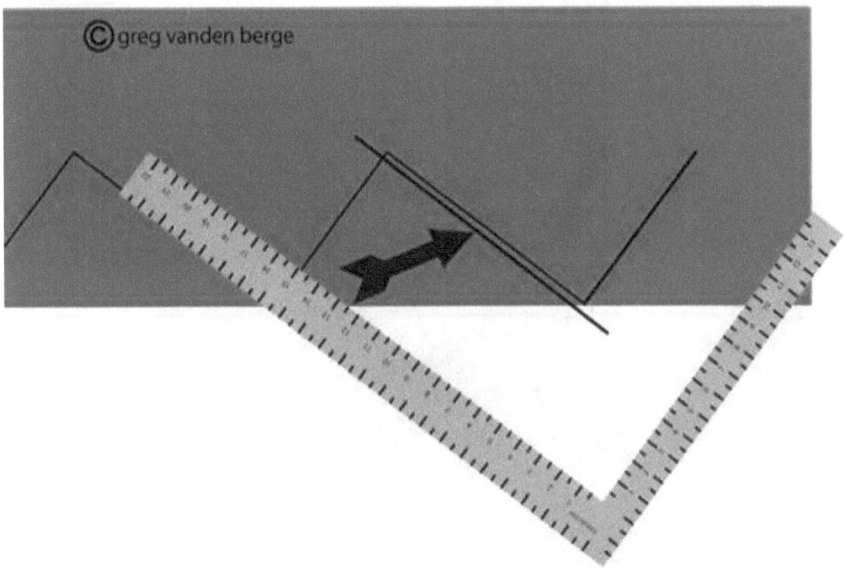

Another thing I would like to point out is that we will be dealing with parallel lines also. When laying out the top or bottom of your stair stringer, you will often use parallel lines to either treads or risers.

The illustration above provides you with a parallel line (the line black arrow is pointing to) that could easily be moved down to form the bottom of the stair stringer.

Bottom Layout Riser Variations

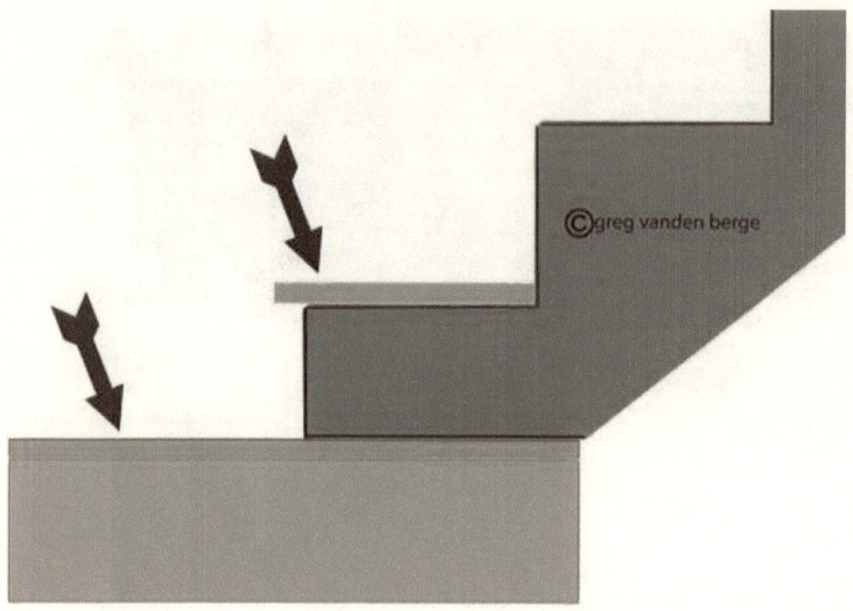

We've already covered this in previous layout methods, but there are additional problems or concerns you should familiarize yourself with. The illustration above shows two arrows, one pointing to a tread and the other, the landing sheathing which are the same thickness.

In a situation like this, all you would need to do is subtract the tread thickness from the individual riser measurement at the bottom of stringer.

If you have a 7 1/2 inch individual riser measurement and a 3/4" plywood stair tread, then all you would need to do is subtract three quarters of an inch from 7 1/2 inches.

The distance from the bottom-top of the stair stringer (top of first step on stair stringer layout, without including your stair tread) to the bottom of the stair stringer would be 6 3/4".

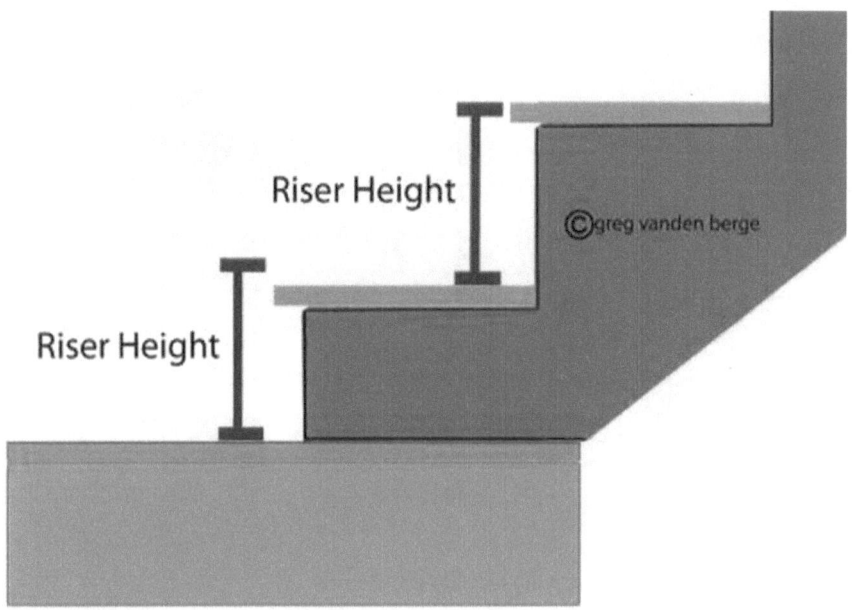

If you subtract too much from the bottom or calculate your measurements incorrectly, you could end up with a situation like this. The first step in the stairway will be lower than the rest and could create a possible trip hazard.

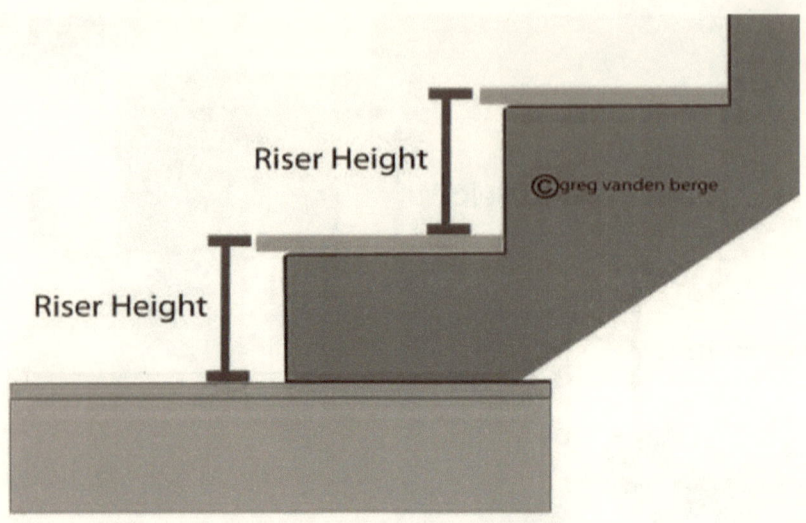

Here's a good example of equal riser heights.

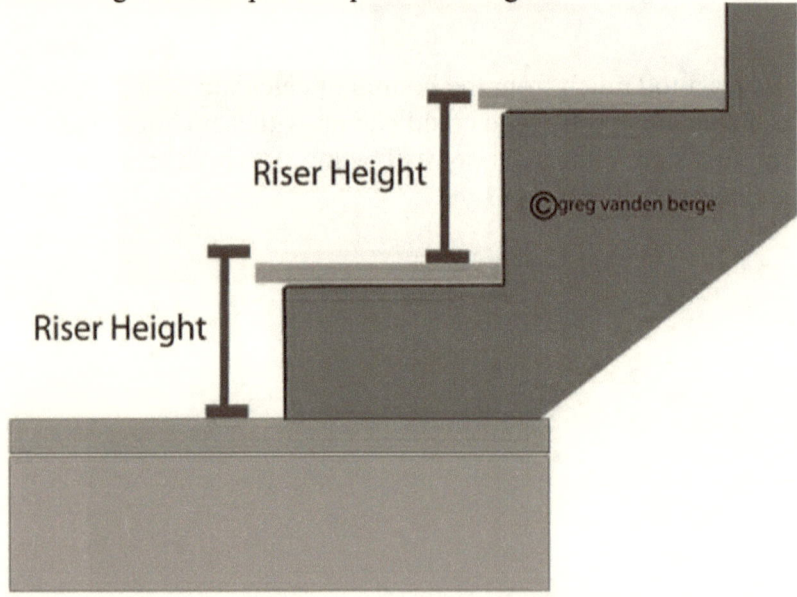

You would think something like this would be a problem, but it wouldn't as long as the top of the landing has been "positioned properly." Stair stringers that sit on top of a landing are usually laid out accordingly.

You can use any materials with different thicknesses as long as you've allow for it. This would apply to any part of the stairway.

I only threw this one in to see if you were paying attention. Sometimes the stuff can be boring, so "WAKE UP!"

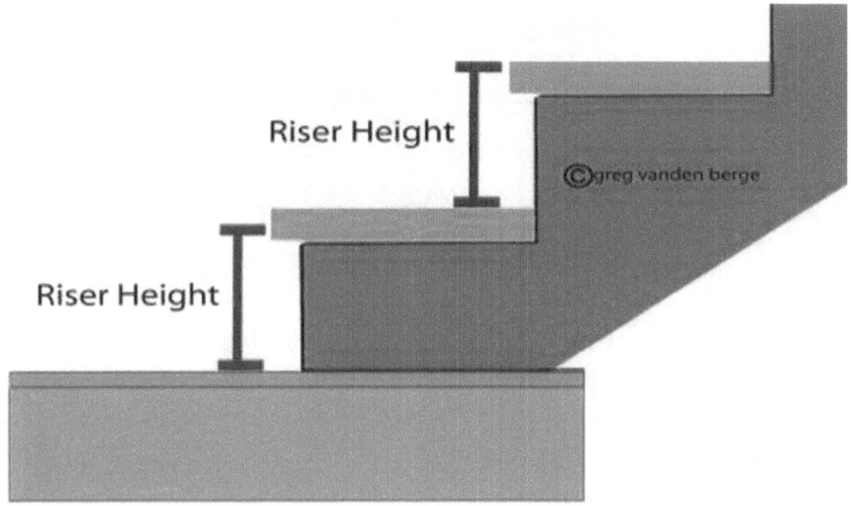

Now this could be a problem. There have been times where I laid out stair stringers, but someone else installed them. If you allowed for a specific stair tread thickness, make sure you share this information with the stairway installer.

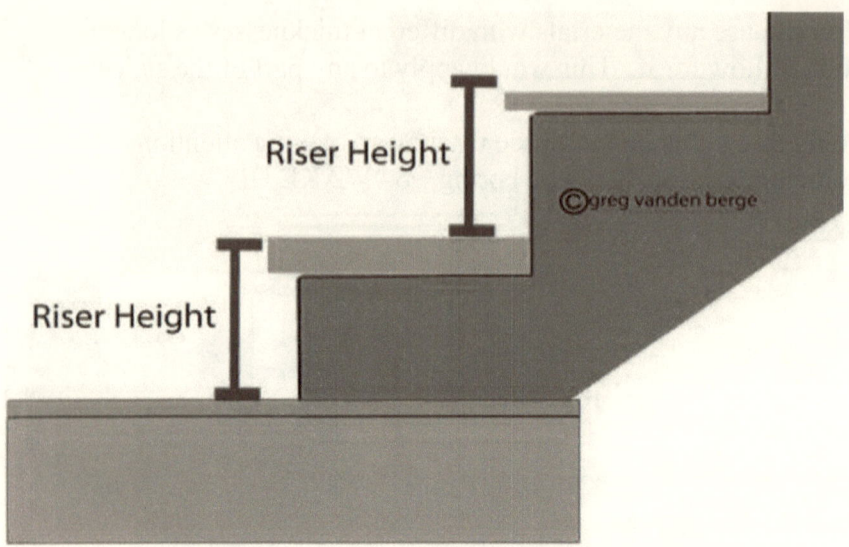

Here's what the stairway could look like if you used different sized stair treads (thickness), without compensating for them correctly. The biggest problem stair builder's face will usually occur at the tops and bottoms of the stairs.

Top Layout Riser Variations

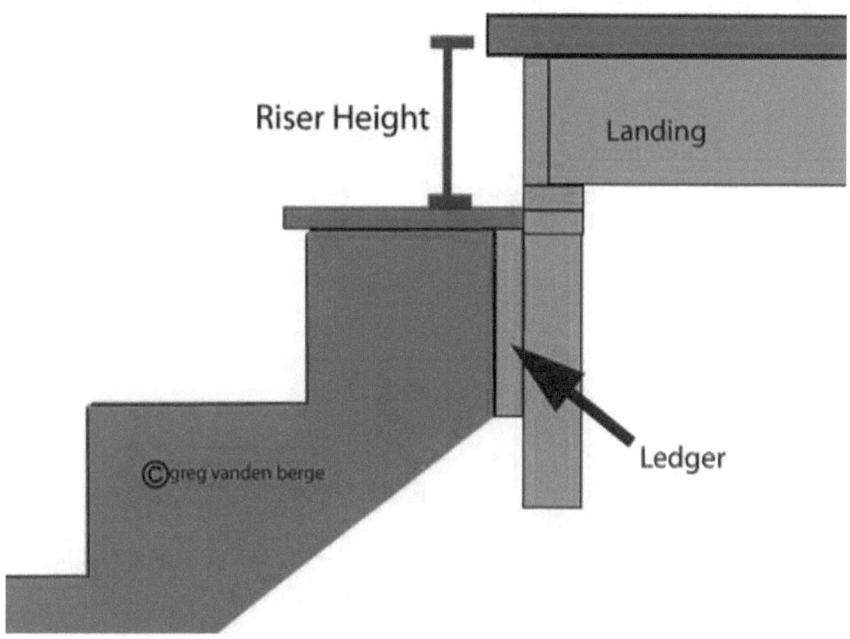

Here's a situation where the ledger should have been moved up to compensate for the landing sheathing thickness. In the picture above the ledger was nailed 7 1/2 inches below the top of the landing joist.

It isn't uncommon to find 1-1/2 inch or 1-1/8 inch thick sheathing or treads, for different parts of the floors or stairway. I've used 1-1/8 plywood for stair treads before, while dealing with 3/4" thick plywood landings and floors.

If you don't get anything else out of this book, pay attention to do this part. I can't tell you how many times I run across a set of stairs like this, built by an experienced carpenter.

Here's a side view of a stairway, with a ledger nailed in the correct position. A situation like this won't affect your stair stringer layout, but will need to be positioned correctly when building the stairway.

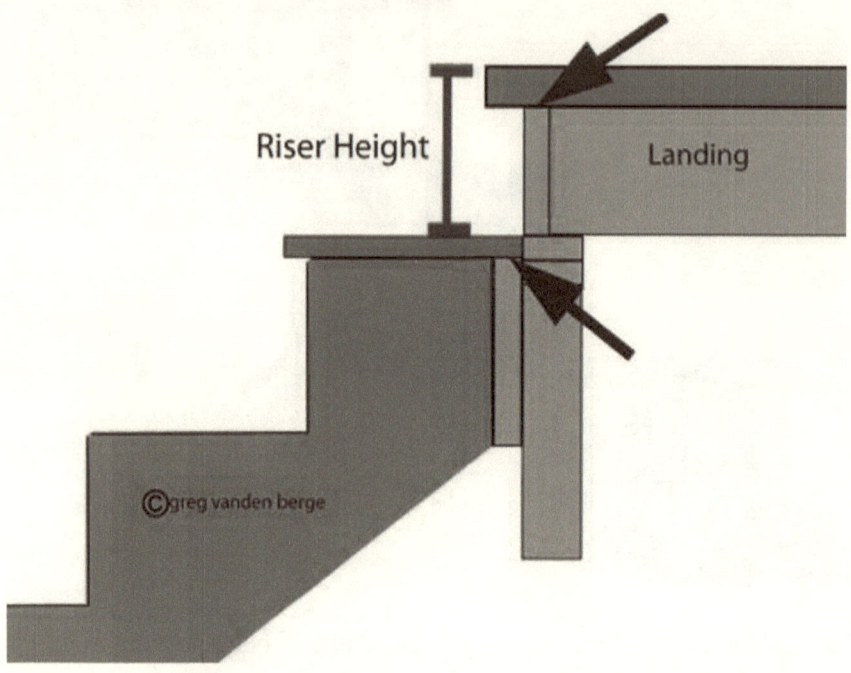

Here's the opposite situation where the stair tread thickness is greater than the landing sheathing thickness. The illustration above should give it away, but if it doesn't, then I will provide you with a little nudge in the right direction. Whenever the tread thickness is greater than the landing sheathing thickness, the ledger will need to be lowered.

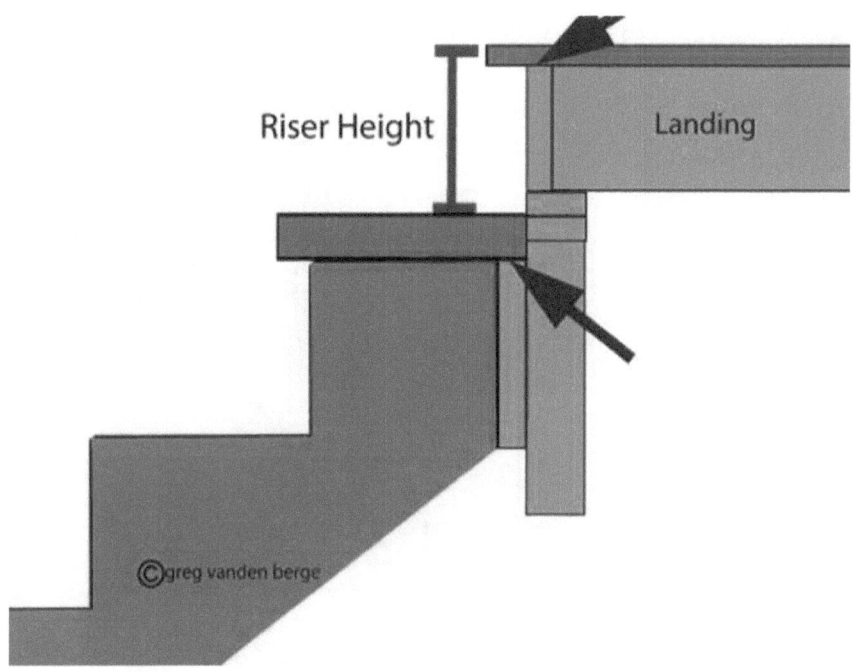

Here's what it would look like when positioned properly.

Riser Variations (Thickness)

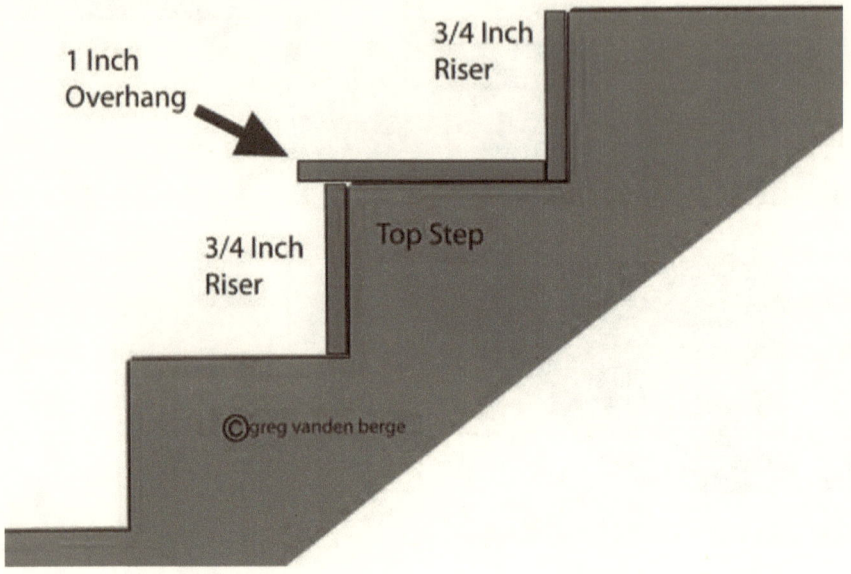

You won't have a problem as long as you're using the same materials for your risers and you lay out your stair stringer accordingly. The illustration above is using 3/4" risers, with a 1 inch stair nosing overhang.

You'll need to compensate for any variations, different sizes or thickness of any and all types of building materials, while laying out your stair stringer.

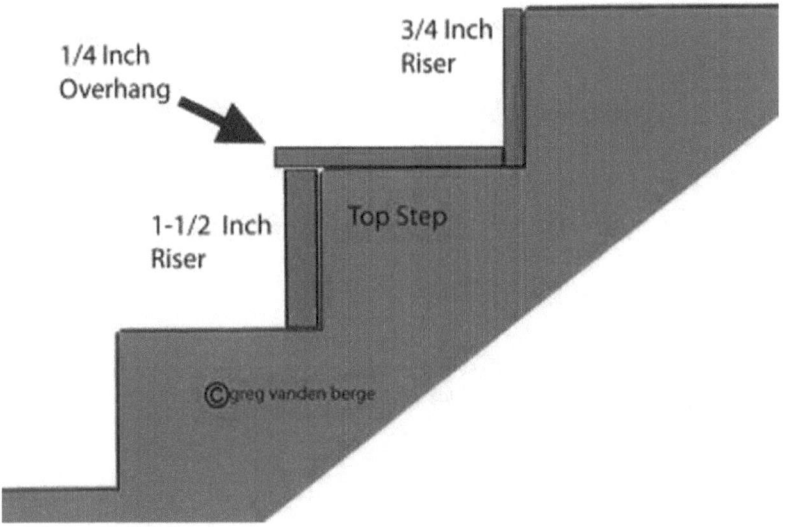

In the illustration above we have an inch and a half thick stair riser on the second riser from the top and a 3/4" stair riser at the top. If the rest of the stair risers are going to be an inch and a half thick, then you would need to add three quarters of an inch to the face of the top riser.

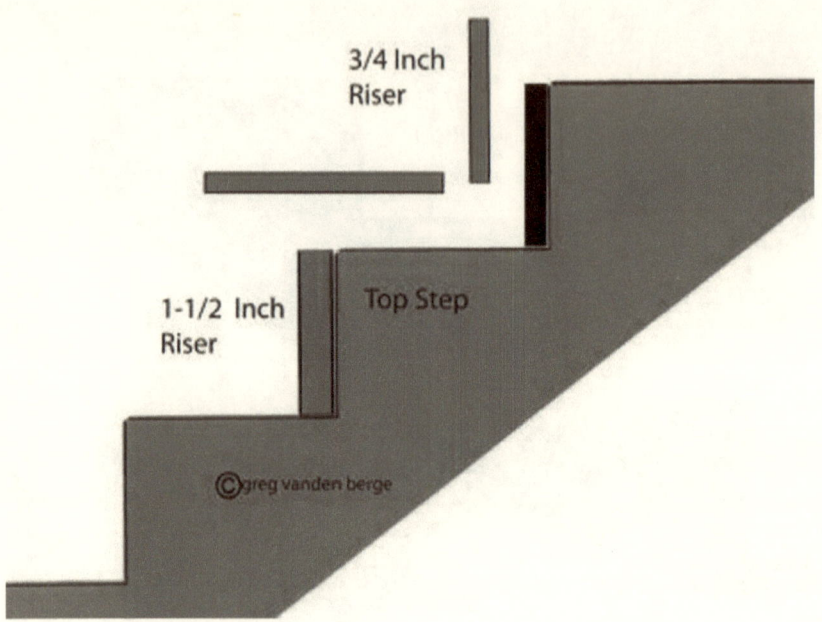

Here's a simple fix, if you ever run into this particular situation. All you need to do is remove the tread and install another riser or other materials to make up the difference.

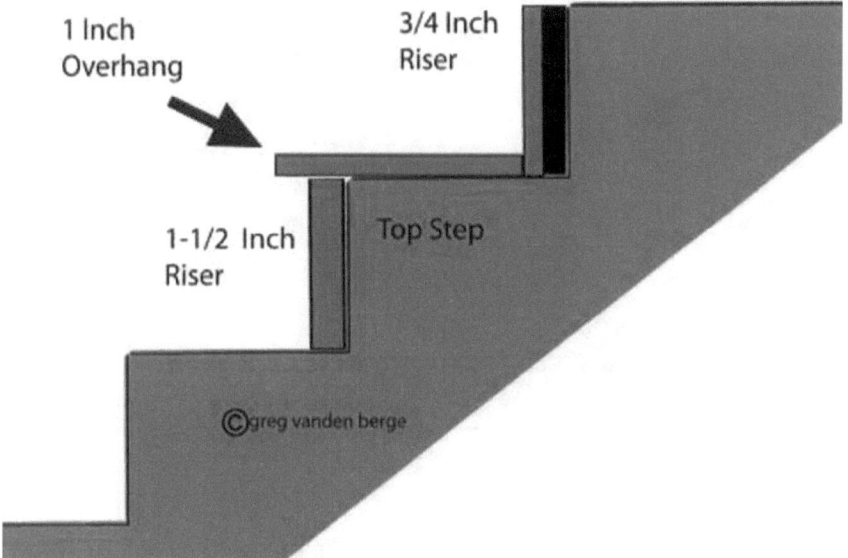

In the illustration above you can see how the modification worked out.

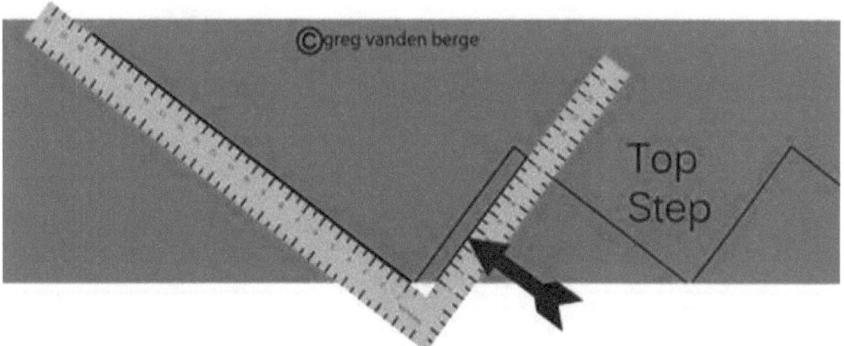

The illustration above provides you with an example of how you can simply add three quarters of an inch to the front of the last riser, in order to make the necessary modifications.

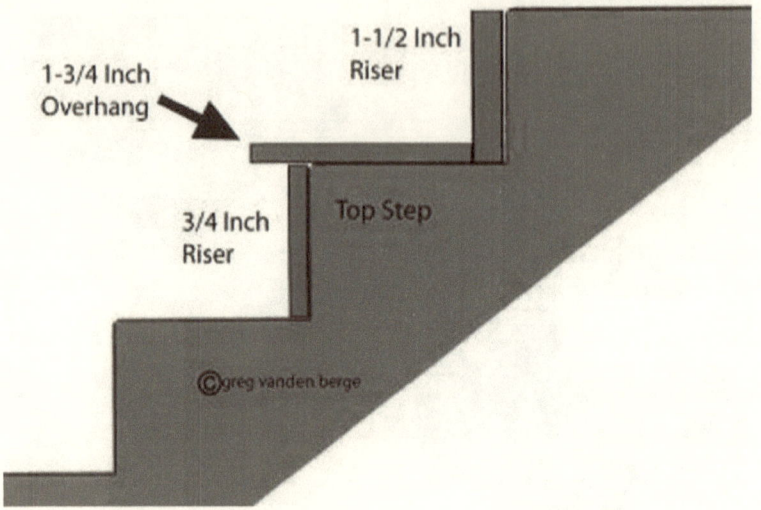

The illustration above provides you with the opposite problem. We now have a 3/4" stair riser, on the second riser from the top and an inch and a half stair riser at the top. This problem will be more difficult to modify while building the stairs, than the previous one.

Hence, the reason why we need to get it right during layout, before you even think about making your first cut, to the stringer.

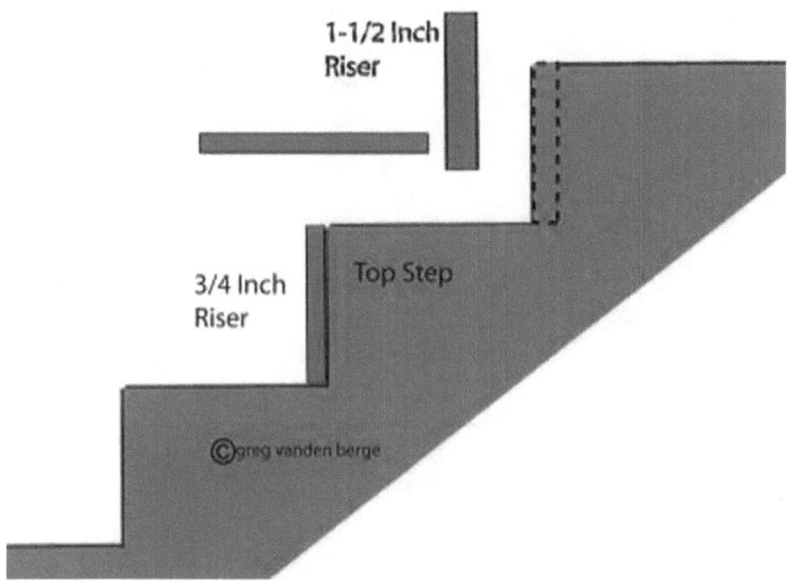

In this situation you'll need to remove three quarters of an inch from the top riser.

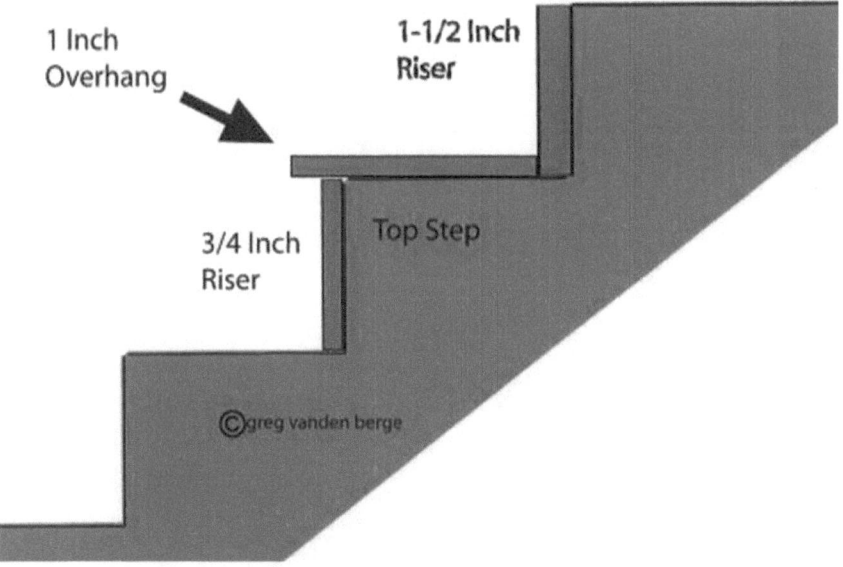

Once we put it all back together and have our 1 inch overhang, along with equal treads and risers, then you've done your job.

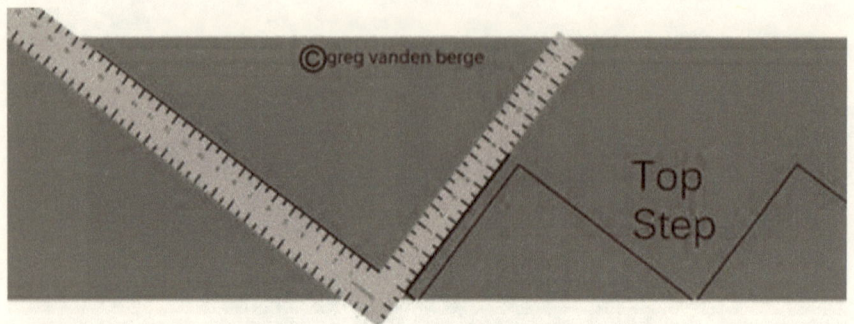

When laying out the top riser, simply slide the framing square back three quarters of an inch, mark the lumber and you're ready to go.

Using Landing Hangers

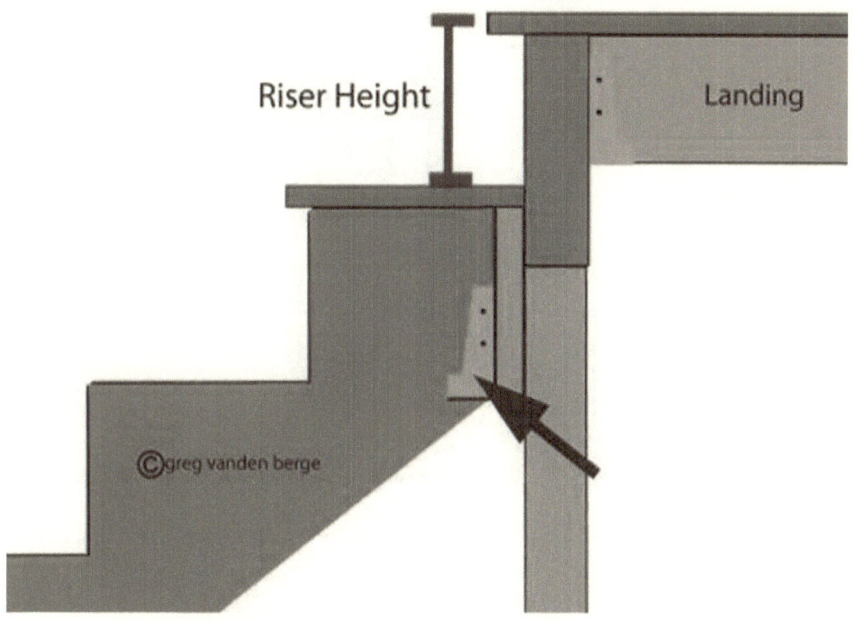

In the illustration above it shows a metal hanger notched in to the back of the stringer. Simply take your saw and cut the desired kerf in the stringer and insert hanger before attaching stringer to ledger.

You won't be able to attach the hanger after the stringer has been nailed to the ledger, for this particular method.

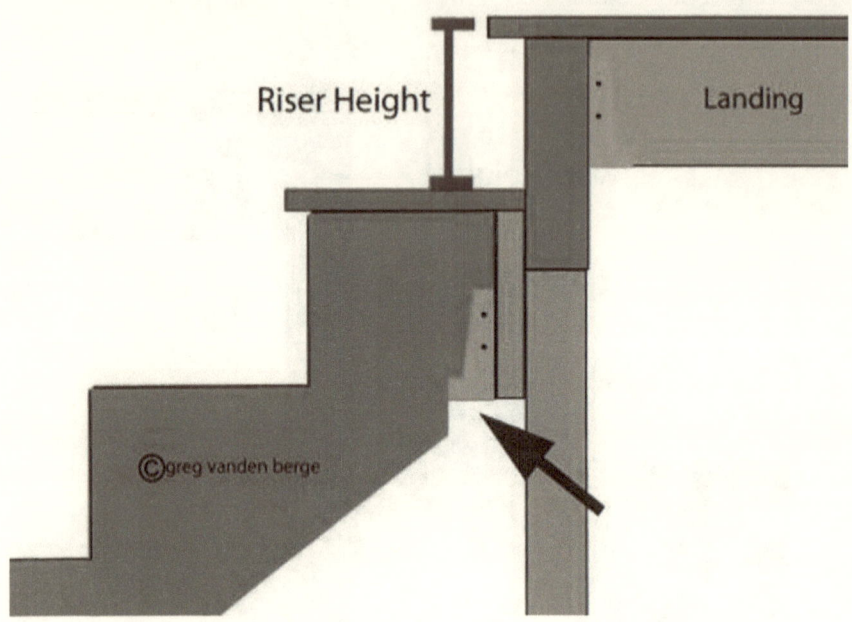

Here's another way to use a hanger, to provide additional structural support for your stairway. Simply notch out enough room under the stair stringer as shown in illustration above, to install hanger.

In this situation, the hangers can be installed after the stringers have been nailed to the ledger, if and only if there isn't anything blocking your ability to nail the hanger off correctly.

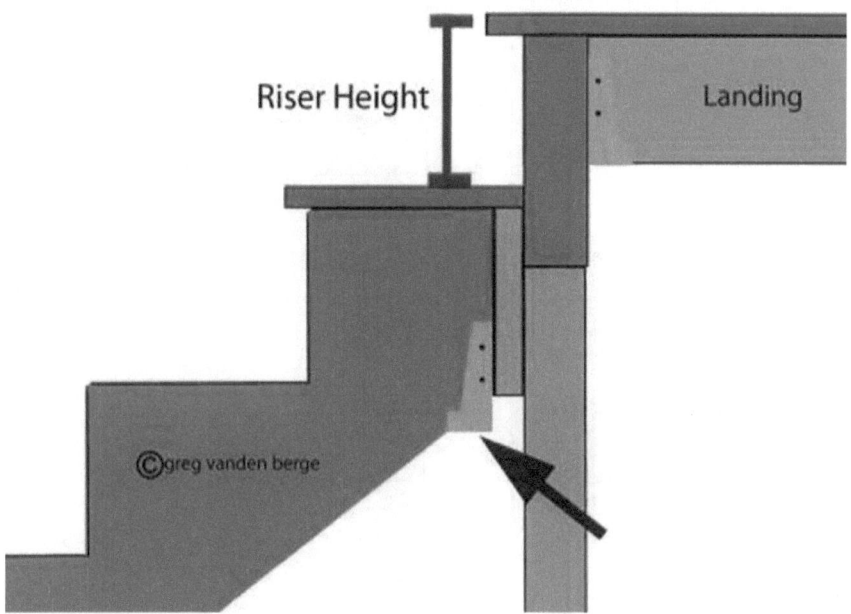

Here's another method that works well. Instead of cutting a notch or kerf into the stringer, simply install the hanger under the stringer as shown in illustration above.

Obviously this method wouldn't work well if any drywall was going to be attached to the bottom of the stair stringer's.

Landing Joist Variations

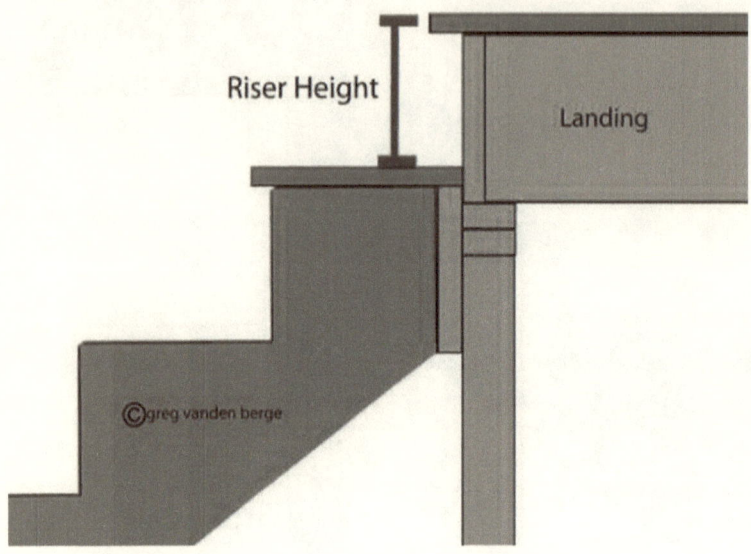

Different sized joist shouldn't affect the stair stringer layout, but you should familiarize yourself with these situations.

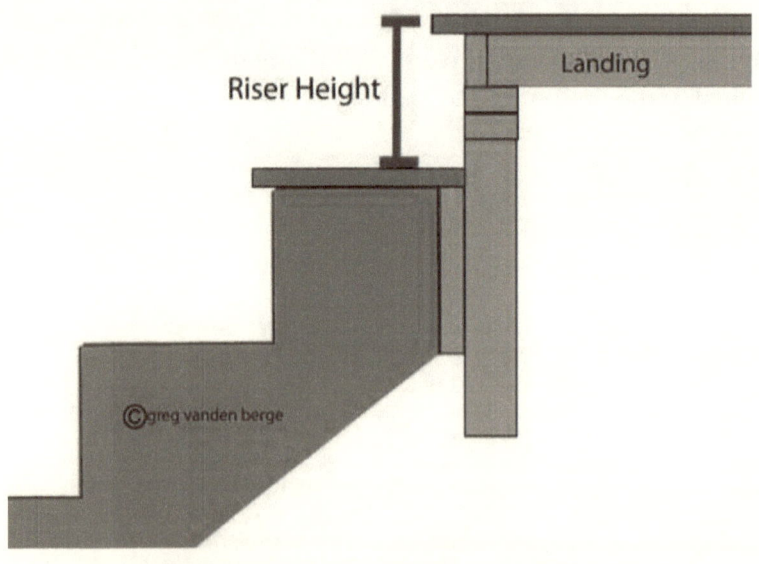

Landing Length Issues

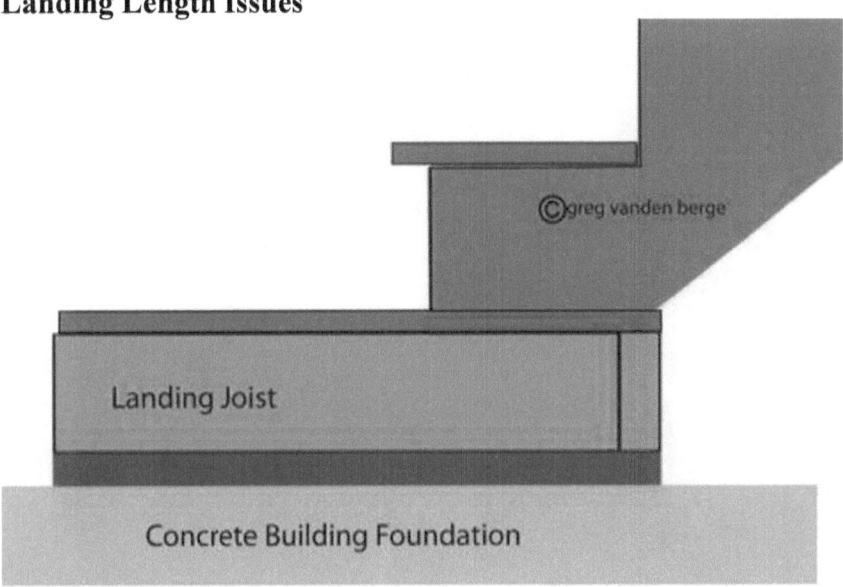

In the illustration above the base of the stair stringer is perfectly positioned on the landing. The landing should support the entire bottom of the stair stringer.

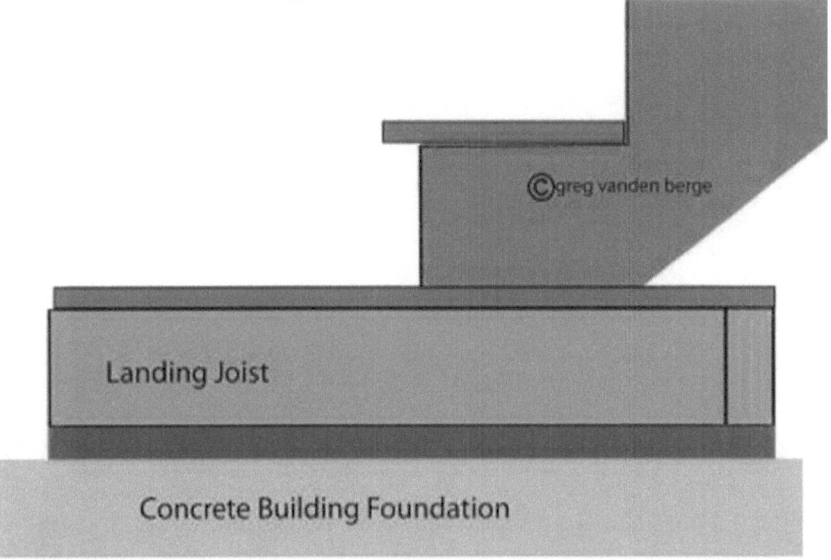

A situation like this would also be acceptable. The stair stringer is still fully supported by the landing.

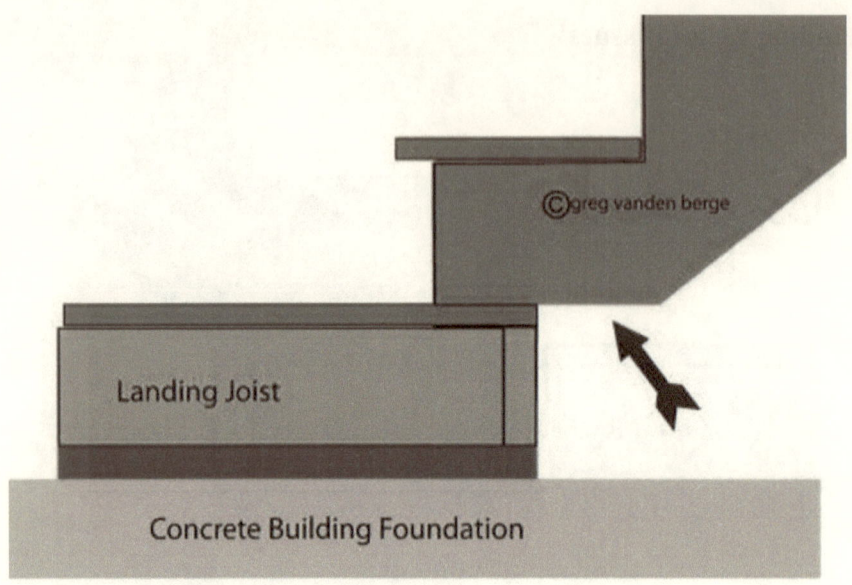

Here's a situation where the stair stringer isn't fully supported by the landing. Now here's the tricky part, I've seen plenty of stairways constructed like this, last for decades.

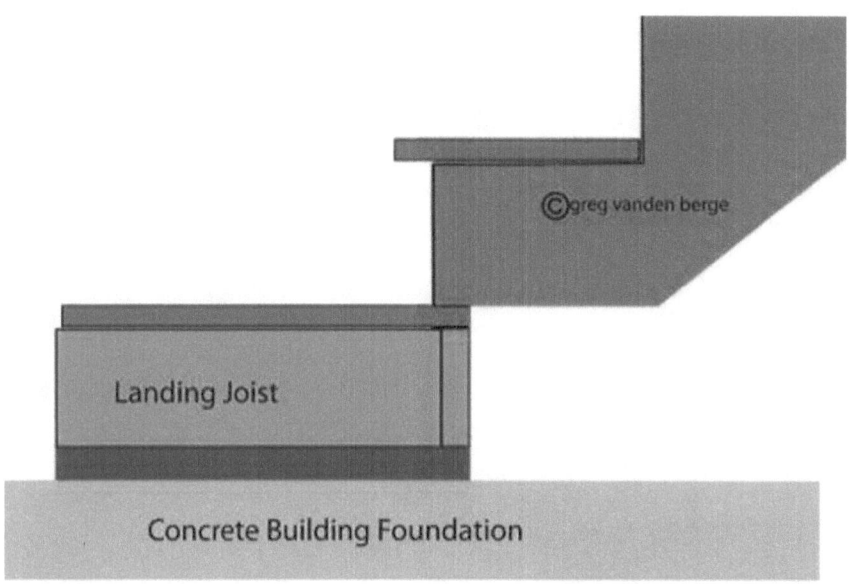

Now, something like this would just be ridiculous, but I've also ran into similar situations and couldn't believe my eyes. It's extremely important that the stair stringer is supported correctly at the top as well as the bottom.

Even in a situation like this, the stair stringer could have been laid out differently, to provide the necessary structural support. It's your job as a stair builder, to make sure the stairs you're planning on building, will last for decades while limiting as many possibilities as you can for structural failure.

Stinger Layout Mistakes

Stringer Cracks And Hangers

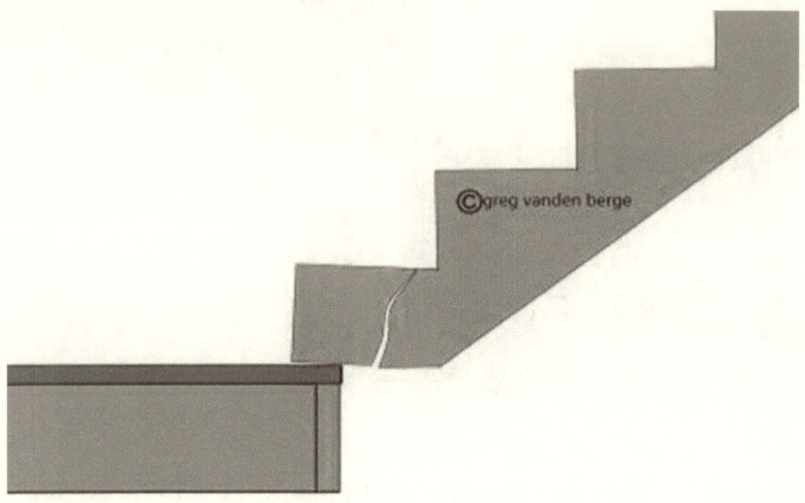

If there isn't anything under your stair stringer to support it correctly then you could end up with a situation like this. It doesn't happen often, but when it does and it results in a personal injury and you're either the property owner or the stair builder, it could be financially painful.

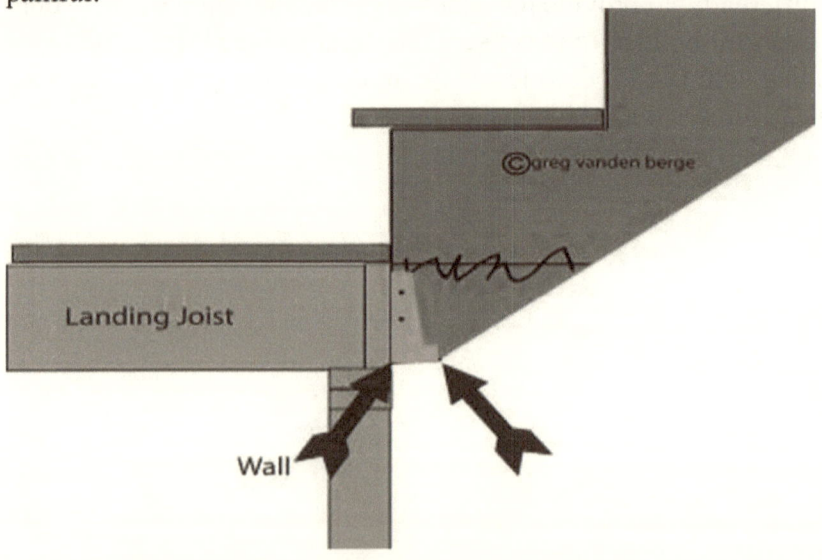

In this illustration the hanger supports the stairway stringer and everything has been laid out and positioned properly.

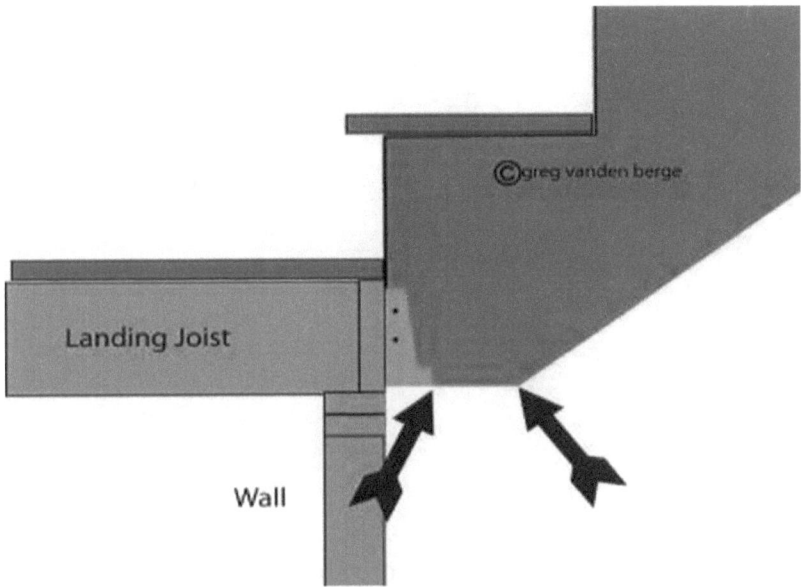

In this illustration it hasn't been. Something like this could last for decades and there have been plenty of times where something like this wasn't a problem. However, you can clearly see the bottom of the stringer isn't fully supported between the arrows.

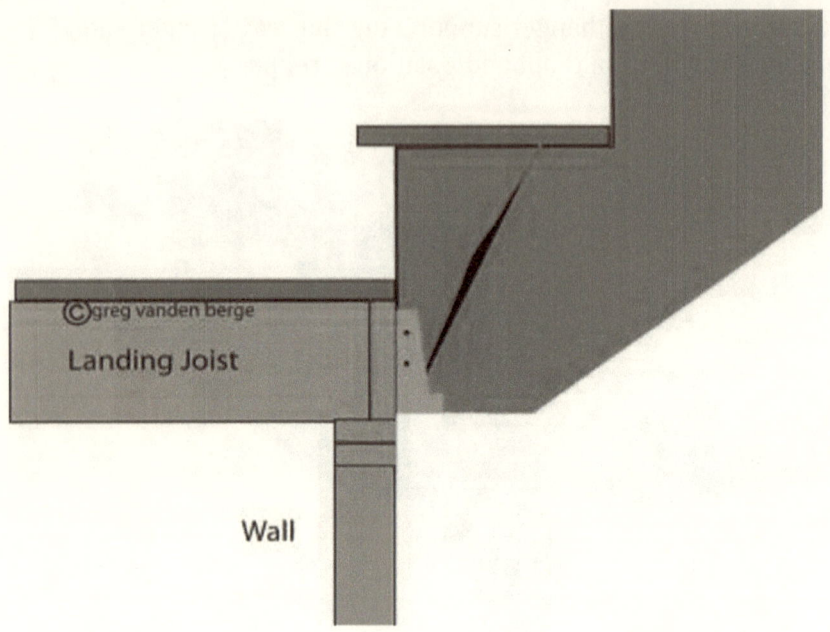

A crack like the one shown in the picture above might not ever become a problem, because it remains supported by the hanger.

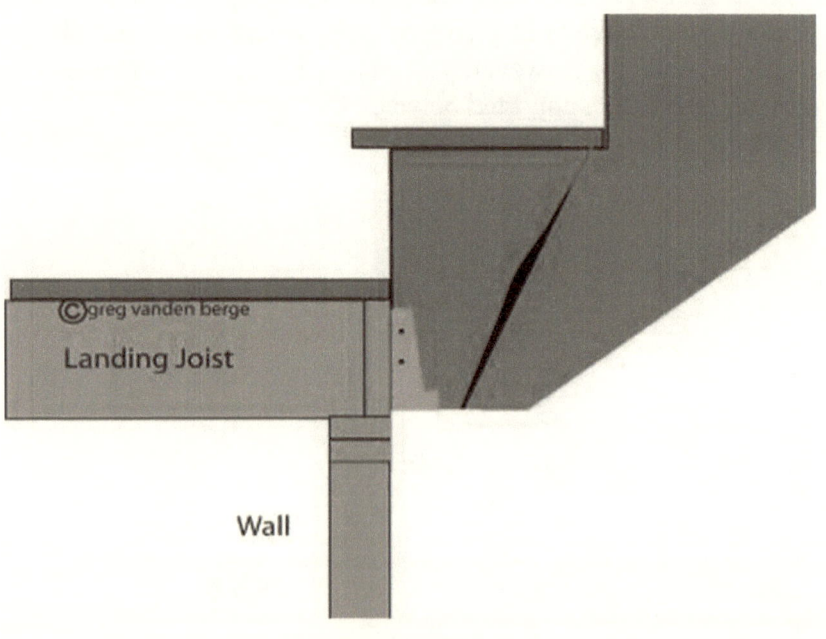

However a crack positioned outside of the hanger or in an unsupported area could be. It's only a matter of time before a stair stringer with a crack like this weakens and breaks.

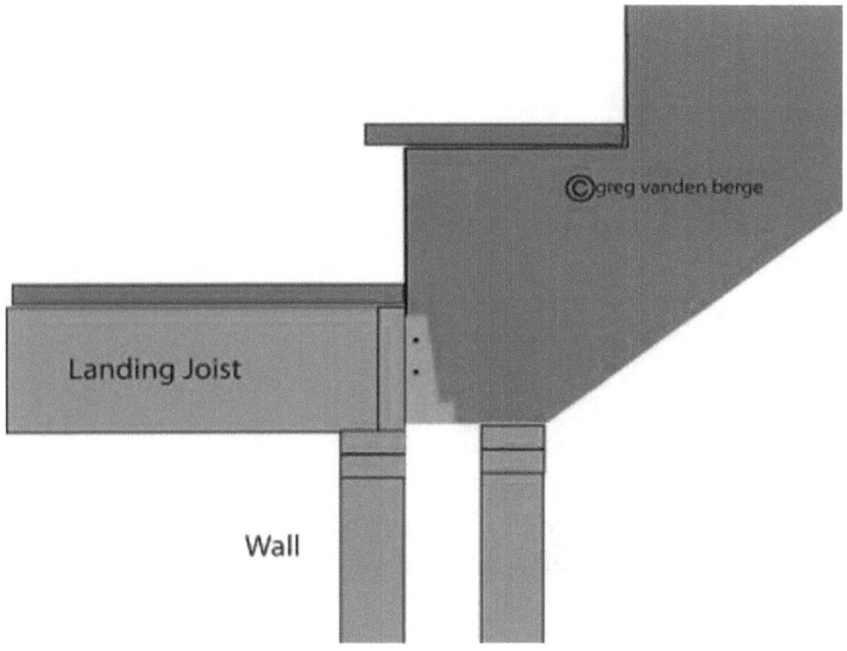

If you do end up with a situation, where part of the stair stringer isn't supported structurally, then you could always add a wall. The illustration above provides you with an excellent example of what I'm talking about.

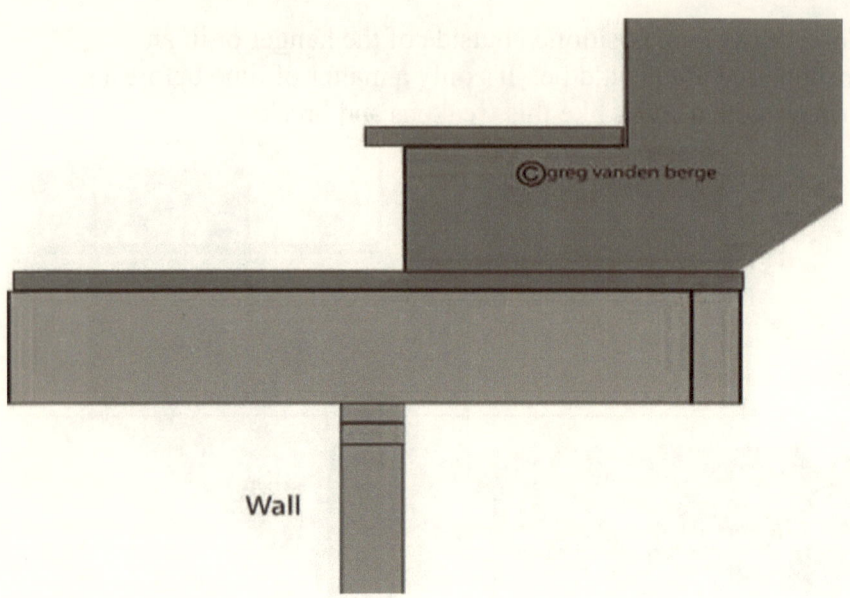

If the stairway hasn't been built yet, you could simply redesign it and lay your stringers out according. The illustration above provides you with another way you could eliminate a structural nightmare, with a little planning and preparation.

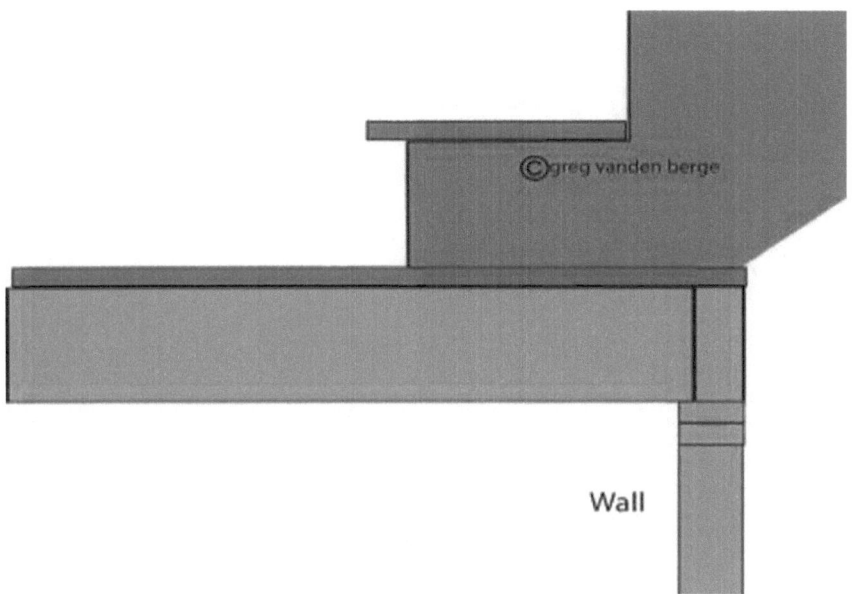

Here's another example and sometimes you'll need to get a little creative, while designing, laying out and building your stairs.

If you're planning on building lots of stairs in the future or just desire to understand the nuts and bolts of how these things are put together, then you'll need to develop an ability to build things in your head.

I can look at a completed stairway and tell you how it was built.

I can also look at a set of plans and picture the entire construction process.

If you can develop the ability to construct stairways in your mind, then you won't have any problems laying everything out and making all of the necessary adjustments, before you ever nail one piece of lumber together.

The End

www.ingramcontent.com/pod-product-compliance
Lightning Source LLC
Chambersburg PA
CBHW030856180526
45163CB00004B/1598